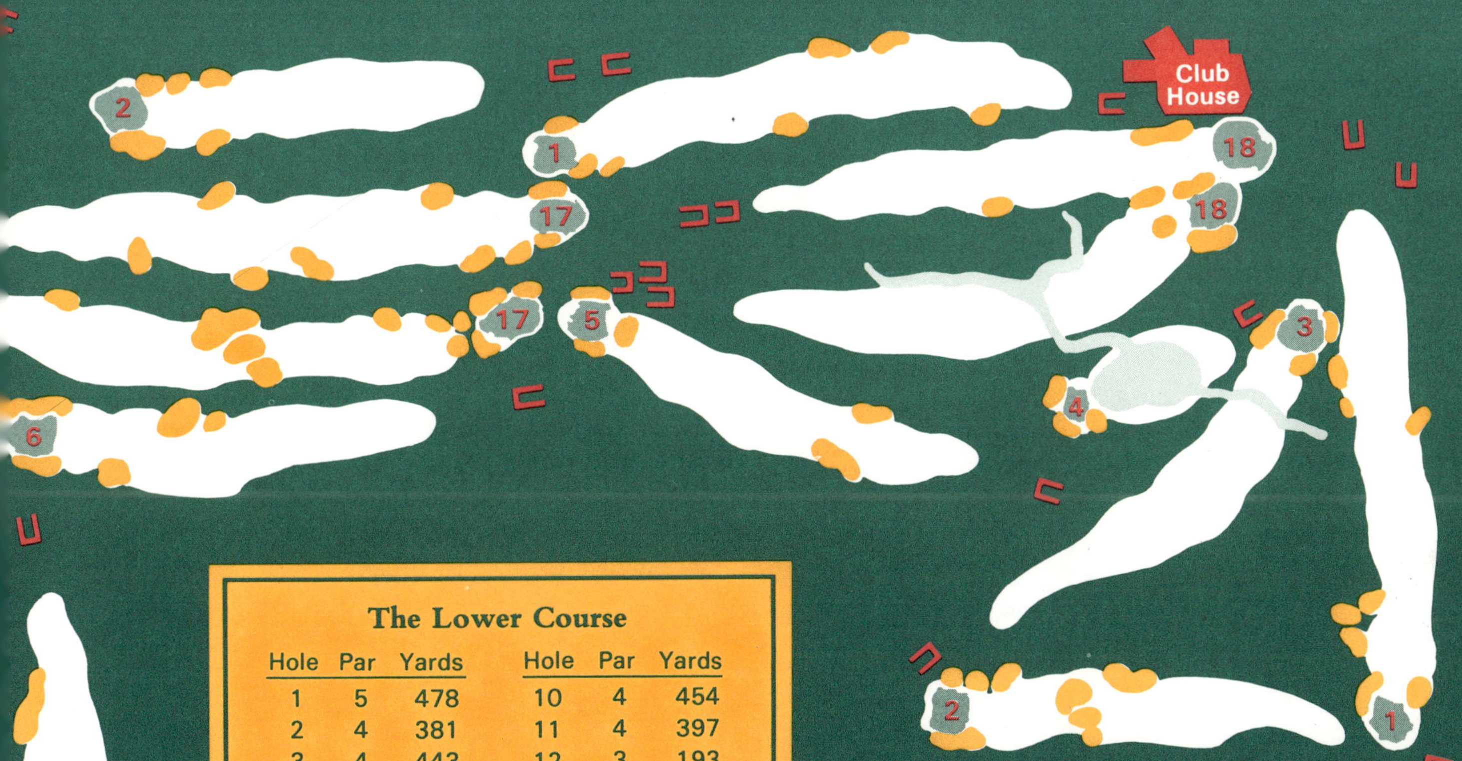

The Lower Course

Hole	Par	Yards	Hole	Par	Yards
1	5	478	10	4	454
2	4	381	11	4	397
3	4	443	12	3	193
4	3	192	13	4	393
5	4	393	14	4	409
6	4	424	15	4	430
7	5	505	16	3	216
8	4	374	17	5	579
9	3	205	18	5	538
Out	36	3395	In	36	3609
			Total	72	7004

BALTUSROL

To
Jim Hand

In appreciation for all you
do for golf, the USGA
+ for being a big supporter
of Baltusrol. You are
such a fine person

Sincerely
Carol Hannon
Pres.

BALTUSROL

90 Years in The Mainstream of American Golf

by James J. Mahon

Designed by Allan Mogel

Ninety percent of the photographs used in this book are
from club files. Several additional outdoor color photos
were taken by Melchior DiGiacomo and others were furnished
by Celanese Corporation. Club interior color photos
furnished by Ariston Interior Designers, Inc. and Walter
M. Ballard Corp.

Endpaper art by Joan Vilato

Prepared and produced by Rutledge Books,
a division of Sammis Publishing Corporation.

Library of Congress Cataloging in Publication Data

Mahon, James Joseph, 1912-
    Baltusrol: ninety years in the mainstream of
American golf.

    Includes index.
    1. Baltusrol Golf Club — History.   I. Title.
GV969.B3M34   1985   796.352'06'0749   85-10707
ISBN 0-87469-049-8

Preface

You might think that Baltusrol just happened — that its history is simply a series of unrelated events that accidentally culminated in one of the world's most prestigious golf clubs.

Of course, this is not so.

Baltusrol is the product of the determined efforts of Louis Keller and several other real-life people who became devoted to a game that was imported from Scotland in the 1880s. They selected a site of great natural beauty on which they built challenging golf courses and an impressive clubhouse. And this small group attracted others of accomplishment and substance who were similarly devoted.

Keller and the other founders were good organizers and managers, and they were followed on the board of governors by an unbroken series of similarly effective people who kept the club on an even keel through two world wars, a major depression, and other disasters both major and minor. Many were captains of industry and finance who gave generously of their time and talents. The professionals they engaged to run the club, particularly "Mr. Outside" and "Mr. Inside," were similarly able people.

The founders were determined to build the best golf club in America. But they also acknowledged and fulfilled their responsibility to the sport in general by helping to form state, regional, and national golf organizations and by hosting as many national tournaments as any other club, and more than most.

In sum, Baltusrol entered the mainstream of golf almost immediately and never left it. Its history is, in fact, a microcosm of the history of American golf.

Len Elliott, well-known New Jersey sportswriter, began work on this history in 1978. He did an immense amount of research, especially on the club's early years. He wrote the Prologue on the Battle of Springfield and the Interlude on the Baltus Roll murder, and he drafted the first few pages of text about the founders of the club.

But death interrupted his effort, and his notes were turned over to John Swart, a former member of the board of governors, for safekeeping.

I have utilized all of Len's research after testing some portions and having others reviewed. Tom Fleming, for example, author of *The Forgotten Victory*, reviewed the Springfield Battle story. And I did the balance of the research and wrote most of the text — except for those portions that were lifted lock, stock, and barrel from the programs of the 1967 and 1980 USGA Opens, which were held at Baltusrol. These portions are the stories of the national tournaments hosted by the club and of Albert W. Tillinghast and the two championship golf courses he designed for Baltusrol, as well as a sensitive portrait of Johnny Farrell. They were written by the well-known sportswriters whose names appear under the stories.

Nancy Rogers and Ruth Swart, both former club champions and women's golf chairpersons at Baltusrol, assembled the information on women's golf at the club.

Carl Jehlen, the club manager from 1953 to 1981, was a valuable source of information about the club's middle years, and Mark De Noble, the present club manager, furnished data concerning recent periods.

Ernest Mudd, maître d' since 1953, and William Eldridge, cocktail waiter, who has been on deck since 1946, furnished much of the material on distinguished visitors to the club.

The manuscript was reviewed and important suggestions were made by Paul Hanna, president of the club; Robert Potter, vice president; James DeBois, chairman of the legal committee; James Collins, chairman of the communications committee; and by Johnny Farrell, the club's long time pro and his son, Johnny Farrell, Jr., a former member of the board of governors. It was Messrs. Jim Collins and Potter who persuaded me to take on the writing project presumably because of my extensive knowledge of the club — I was a governor from 1969 to 1977 — and my authorship of other published works. I was skeptical though about what I would find.

Finally, the manuscript was reviewed by my octogenerian friend and fellow club member, Joseph Partenheimer, in whom I have a special confidence.

All of this is to express a fervent thanks to those who helped with the work — not to shift the responsibility for its contents which, of course, must remain mine.

I am now willing to admit that Baltusrol has a history and an interesting one at that. I hope my fellow club members think so too.

James J. Mahon
Short Hills, New Jesey
November 1, 1984

CONTENTS

Prologue — An Historical Site   9
Interlude *or* Murder Most Foul   11
Chronological Highlights   13
Conception and Birth: Louis Keller (1895–1898)
Coming of Age: Keller–Bayard (1898–1919)
The Calm Twenties: Sinclair–McKnight (1919–1929)
The Depressing Thirties and the Tight Forties: Baker–
Brown and Major Jones (1929–1943)
Restoration: The Steady Ones and Carl Jehlen
(1943–1958)
Improvement: The Venturesome Quintet (1958–1973)
Expansion: The Merchandisers (1973–1985)
The Anchors to Windward

The Golf Courses   43
The Old Course
The New Courses

Contemporary Scenes: A Color Portfolio   65

National Championships Played at Baltusrol   97

**Metropolitan and New Jersey Championships
Played at Baltusrol**   117
Baltusrol Golfers   121
Men Golfers
Women Golfers

Club Professionals   135
Willie Anderson
George Low
The Legendary Johnny Farrell
Bob Ross

Other Sports and Activities   142
Members and Guests   144
Indecision and Headaches   147
Appendix   151
Presidents
Club Champions
Seniors Association Champions
Outside Championships
Women Champions
Women's Golf Chairpersons
USGA Championships Held at Baltusrol

Index   157

Prologue—An Historical Site

England's war with the American colonies was not going well for King George III. Burgoyne's thrust down the Hudson Valley to cut off New England had ended in disaster at Saratoga. The French had come in on the side of the colonists, and the English retreated to New York. General Washington and the determined Samuel Adams were showing no signs of giving up. And most of Europe, as King George lamented, was against him.

Although the King knew nothing about it, communications being what they were at that time, a move was launched in the spring of 1780 from Elizabethtown, New Jersey, toward the Continental Army in Morristown. The aim was to rout Washington's army there, knock New Jersey out of the war, and perhaps even capture Washington himself.

The commander in chief of the area was General Sir Henry Clinton, but the move, when it was made, was not his idea. Sir Henry had other notions about what should be done, but he was off in South Carolina, capturing Charleston from the Americans.

The attack was made by Hessian General Wilhelm von Knyphausen, who was in command during Sir Henry's absence. Actually, it was not his idea either. He was under severe pressure from a group of Tories in New York who were chafing at British inaction. Chief among this group were William Smith, an aristocrat, who was banished from the state when he refused to take an oath abandoning loyalty to the King, and William Franklin, illegitimate son of Benjamin Franklin.

They and their group were getting a flood of intelligence from spies in New Jersey, all to the effect that the people there were ready to go over to the British if given half a chance. The militia on whom Washington had to rely would not fight, the Tories had been told, and the Continental Army in Morristown was in bad shape.

The time was perfect, the Tories insisted, for a quick push from Elizabethtown through both Springfield and the Hobart Gap to Morristown.

The Tories were so insistent that von Knyphausen finally gave in, partly persuaded by the intelligence that the New Jersey colony was ready to give up and perhaps scenting some glory for himself if his thrust brought about the collapse.

So, on the night of June 6, 1780, von Knyphausen ferried a force from Staten Island across what is now the Arthur Kill to Elizabethtown Point. The main British and Hessian force was to follow in the morning. This move got off to a bad start when a group of 4th New Jersey Regulars warned by an American spy that the British were coming, fired a volley in pitch darkness that seriously wounded General Thomas Stirling, the best officer on the British side.

Von Knyphausen pressed on anyway, only to be stopped at what is now the town of Union by the resistance of the American regulars and the militia that the British had been told would not fight. When von Knyphausen learned that Sir Henry Clinton was soon to arrive from Charleston with other plans for the New Jersey campaign, he drew back to Elizabethtown Point.

Sir Henry did arrive, and when he learned what had happened, he was furious. All they had done, he said, was to alert Washington, the Continental Army, the militia, and the whole Jersey countryside. Notwithstanding, Sir Henry went ahead and organized a full-bodied invasion.

The assault, launched on June 23, swept through Elizabethtown, up what is now Morris Avenue (or Route 82), and through Union with hardly a pause. A second prong came up Vauxhall Road and took Vauxhall.

In spite of sporadic defense by the militia, who were expected to lay down their arms, the attacks went well until the left column reached the Rahway River at Springfield. There, the troops ran into the Continental Army led by General Nathanael Greene and bolstered by the militia. The defenders exacted a heavy toll of British and Hessian troops, although the river crossing was finally forced. The invaders were halted only when they reached the west branch of the Rahway, just past the Presbyterian Church at what is now Springfield Corners. Parson Caldwell tore up psalm books during the fight and tossed them to American soldiers for cannon wadding, crying, "Put wads into 'em, boys!"

The Vauxhall prong, meanwhile, crossed both branches of the river and reached almost to the old

Springfield Church

paper mill in Millburn. There, they were in a bad position at the foot of what was then known as the Short Hills.

But by now it was noon and von Knyphausen paused to rest and take stock of the situation.

The fighting had been bitter, he knew from the number of British casualties, and it had been infused with a special hatred because some British detachments were made up of American volunteers still loyal to the Crown. Neighbors and families had been split by their sympathies. The intensity of feeling had been heightened too by repeated raids the British and their American units had made all along the coast, from Newark to the Amboys—raids that involved looting and burning and the killing of non combatants.

The more von Knyphausen looked at things, the less he liked them. His two prongs could possibly have bulldozed their way through the Hobart Gap, but it would have been at a very heavy cost. So the general ordered a retreat to Elizabethtown Point. It was well he did, too, for Washington was on his way from West Point with more of the army, and General Greene was so satisfied with his position that he was hoping the British would attack.

The Americans first knew of the British retreat from the sight of smoke rising from the houses of occupied Springfield. When the British retreated, they usually set fire to as many buildings as they could. Among the burning buildings was the Presbyterian Church. But the church was rebuilt in 1791 and it still stands.

That was the end of the Battle of Springfield, which according to Thomas Fleming in *The Forgotten Victory* was the battle that finally discouraged the British from trying to crush the Revolution by force. The British and their American volunteers made a rapid withdrawal, harassed all the way by the infuriated militia and Continentals, many of whom had friends or relatives in devastated Springfield.

Who could have dreamed then that more than a century later, two British monarchs would enjoy themselves at golf barely a mile from historic Springfield Corners and be welcomed there as honored guests. They were then King Edward VII ("Bertie"), who played at Baltusrol Golf Club in 1907, and former King Edward VIII (the Duke of Windsor), who visited the club many times after his abdication in 1936.

Interlude *or* Murder Most Foul

This inscription is on a marker in the Revolutionary Cemetery in Westfield, New Jersey. It is appropriate, too, for it marks the grave of one Baltus Roll, a Springfield farmer who did indeed meet a sudden death on the night of February 22, 1831. His first name was spelled B_o_ltus on his tombstone but an *a* survived the *o* in other references.

It was a violent period in American life, only sixteen years after the second war with Great Britain. The frontier had been pushed west of the Alleghenies, and clashes with the Indians were frequent and bloody. The President at the time was Andrew Jackson, himself a quick-tempered, violent person, who had killed one man in a duel.

So, although the murder of Baltus Roll was not especially unusual, it was to have a certain significance.

February 22, 1831, was one of those dismal winter days. Snow lay heavily on the ground, but that night it rained hard. Roll, who had a small farm with a house on the mountain between Springfield and Summit, retired early. About midnight his wife was awakened by a pounding on the door and a gruff voice demanding to be let in. She refused, but that did not suffice. The door was battered in, and two men entered the house. According to Mrs. Roll's testimony at the trial of one of these men, they dragged Roll from his bed, "slatted him about the room and hauled him outdoors." She ran out, saw the men tie Baltus and throw him into a puddle of icy water. He called her name twice, she said, but after that "he did not make a sound and I thought he was dead."

When the men went back into the house, Mrs. Roll ran down a path and into the woods. The snow was deep, it was raining, and the wind was blowing hard. She came back but kept out of sight of the house, fearing what the men would do if they caught her. Finally, toward daylight, she went to the home of a neighbor, Jesse Calhoun, and told him what had happened. He got Brook Sayre, who was Roll's cousin, and Joe Cain. Together they went to the Roll home.

What Mrs. Roll had told them was, of course, true. They found Baltus's body by the gate, and the house itself had been thoroughly ransacked. The murderers had emptied all the drawers and closets, torn up the carpets, overturned the beds, and even torn some bricks out of the fireplace. They were looking for money.

Two men, Peter B. Davis and Lycidias Baldwin, were suspected of the crime almost immediately. Davis was known to have been badly in need of

Baltus Roll's house

11

money and to have looked for someone to go with him "where we can get a thousand dollars." Davis was quickly arrested.

When Baldwin heard that the police had Davis, he went to Morristown and killed himself in the room of a tavern.

Davis went on trial in Newark for the murder and, strangely, was acquitted because the judge ruled some of the evidence inadmissible. During the trial he did admit forgery, however, and was later arraigned on four counts. He pleaded guilty to three of them and was sentenced to eight years on each one.

Davis died in jail.

Although Baltus Roll certainly was an obscure figure, the murder and trial became a newspaper sensation in the New York metropolitan area. Why, no one seemed to know. Somehow it caught the imagination of editors and, as we say now, it took off.

So Baltus Roll went to his final resting place unaware that his name, in a slightly contracted form, would be immortalized in a mountain, several streets, and a golf club known nationwide and even across the seas.

CHRONOLOGICAL HIGHLIGHTS

Conception and Birth: Louis Keller (1895–1898)

The year was 1895. Theodore Roosevelt was trying to clean up the New York City police department. Grover Cleveland was in the third year of his second term as President and was disputing with the British over a boundary in Venezuela. The French had about given up on cutting a canal through the Isthmus of Panama, and some Americans were saying we should take over the job. The National Baseball League was firmly established and dominated by John McGraw and the Baltimore Orioles. Wall Street and the country were recovering from the Panic of 1893.

Such events identify the period, but they were of little interest to a man who lived in New York in those so-called Gay Nineties, one Louis Keller. His thoughts were elsewhere. Keller had been a member of the Short Hills Club, a racquet club, since at least 1881. According to the club's centennial history, published in 1975, he had agitated for the establishment of a new clubhouse and a golf course on farmland in Springfield at the foot of Baltusrol Mountain. Keller was unsuccessful—so he decided to go it alone. Around 1890 he bought up some 500 acres of farmland in Springfield Township and fixed up a house on the property as a country home. He was then living in the Calumet Club on New York's West 56th Street.

In the spring of 1895, Keller sent a two-paragraph letter to selected friends, including fellow members of the Short Hills Club, inviting them to become members of the new golf club he was forming. It was such a simple letter that it is hard now to believe that it eventually had repercussions nationwide. Still, as the Chinese have reminded us, a journey of a thousand miles begins with a single step.

The first paragraph of Keller's letter read:

You are invited to become a member of the Baltusrol Golf Club, the grounds of which are at the foot of Baltusrol Mountain, midway between Orange, Morristown, Newark and Plainfield, about 17 miles from New York. A course of nine holes, averaging about 250 yards and with 40-foot greens, has been laid out upon the sandy hills naturally adapted for the purpose and now is ready for use. An eight-room house on the grounds will be fitted up with grill and clubhouse facilities.

The second paragraph dealt with directions on how to reach the club.

The first step had been taken.

Keller got a prompt and enthusiastic response to his letter, so that by the fall of 1895, October 25 to be exact—just seven years after the organization of the first permanent golf club in the United States, St. Andrews in Yonkers, New York—he sat down in his New York office with three other men and organized the club. The other three were Arthur D. Weeks, John DuFais, and Arthur B. Turnure—with Keller, the club's first board of governors. Weeks was elected president, DuFais vice president, Turnure treasurer, and Keller secretary—a position he would hold until his death in 1922.

All four were born in New York City, Weeks in the old family home, a landmark on Bleecker Street in Greenwich Village. (The name then was spelled *Weekes*, but the third *e* was later dropped.) A Columbia University graduate, he became an expert in real estate law. He was long active in the exclusive Union Club and was its vice president at his death. Weeks was also an uncle of Alice Weekes, who startled New York society by appearing as a ballet dancer in the musical *Aphrodite*, at a time when young women of her social station were not supposed to do such things. She used an assumed name, of course, but society was not fooled.

Weeks was not a man to be intimidated, either. In 1920 he raised the Union Jack and the Tricolor at the Union Club, honoring the 300th anniversary of the landing of the Pilgrims and the return of the French ambassador. An anti-British mob stoned the building, but Weeks refused to lower the flag, "coolly ordering it to remain unfurled," *The New York Times* put it. Living until 1925 and the age of 75, he saw the two Baltusrol courses that are in play today.

John DuFais was an architect out of Harvard and MIT. He worked with various New York firms on remodeling jobs, such as those of the Vanderbilt and Goelet homes and the Capitol building in Albany before setting up his own firm. He designed the Union Club in New York, among other buildings, and worked with the renowned Cass Gilbert on some remodeling of the White House. DuFais wore a beard and a large mustache. One of the early American golfers, he was present at the birth of the Metropolitan Golf Association at Delmonico's in 1897 and was its first secretary. He died at the age of 80 in Newport, Rhode Island.

Arthur Turnure graduated from Princeton in 1876. He held a law degree but did not practice, being much more attracted to art. He published two books on art, was art director of Harper & Bros., turned out a richly illustrated edition of *Ben Hur,* and eventually left Harper to publish *Vogue,* of which he became president. Turnure organized the Grolier Club in 1890, reorganized the Society for the Prevention of Cruelty to Animals, and was prominent in women's charities. Mrs. Arthur Turnure was runner-up to Beatrix Hoyt in the 1896 U.S. Women's Championship held at Morris County Golf Club. Arthur Turnure died in 1906, at the age of 49.

And what of Louis Keller, the man who started it all?

Keller was born in 1857, the son of Charles M. Keller and Heloise de Chazournes. His father was a patent lawyer and at one time Commissioner of U.S. Patents. In his twenties, Louis tried his hand at several things — gunsmithing, dairying, even organizing the American Wheelmen. None worked out very well. But in 1887 Keller hit the right key. He founded the *Social Register* of New York. The *Register* immediately became popular, probably because Keller avoided the pitfalls of interviewing candidates for inclusion in the *Register* or publishing the ground rules. Rather, he apparently set himself up as the sole arbiter of social acceptance. His assumption of the arbiter role may not have met with everyone's approval, for Cleveland Amory in *Who Killed Society?* quotes a fellow club member as saying that Keller was "a gentleman — but just."

Keller was described as rather short, with sandy hair and a drooping mustache. He had been somewhat deaf since childhood. According to John C. Smaltz, who joined Baltusrol in 1920, Keller did not have a highly developed sense of humor and was somewhat of a loner. He was seen most frequently lopping off branches from small trees and shrubbery around the clubhouse.

Ironically, the Baltusrol Golf Club was founded by a man who did not play golf. He had many acquaintances who did, however, and he noticed that

Old Clubhouse with third addition

they played in Newport, on Long Island, or in Westchester County in the summer, but in the winter they went to Florida or to some other warm climate.

Observing the southeasterly exposure of his land at the foot of the mountain (from which the World Trade Center can now be seen), Keller decided that a golf course there would enable the addicts of this new game to play all year round. He thought this would be especially attractive because of what he termed the "mild winters" in New Jersey—presumably ignoring the blizzard of '88.

In 1922 Charles Phelps Cushing said in *Country Life* that Keller had George Hunter, an Englishman, lay out the first nine holes referred to in his letter of invitation.

It is also worth noting that in his letter Keller already had given the club the name it has today. A clue to this is contained in an 1895 issue of the *New York Herald*. In a story about the formation of the club, Louise McAllister is named. She, it said, "is believed to have chosen the name." Louise McAllister was the daughter of Ward McAllister, a Californian who came east and quickly became the czar of New York society. Keller, who never married, had little to do with women. Yet, the *Herald* tells us he "gave several fêtes champêtres at his country place in her honor." Miss McAllister later married and as Mrs. Albert Nelson Lewis was one of a dozen people who attended Keller's funeral 27 years later. What could have been more likely then than that Miss McAllister coined the name, running the murdered farmer's first and last names together, Baltus and Roll, and dropping off the final *l*. The word trips easily off the tongue.

Nowhere in Keller's letter of invitation was the cost of membership mentioned. Actually it was $10 and that the original membership numbers close to 50 is borne out by the action of Keller at that October 25 meeting. He turned over $490 to treasurer Turnure as cash on hand.

If you think $10 was an incredibly low figure, consider where we were at that time. The buying power of the dollar in 1895 was many times that of today's inflated dollar. Men's suits were advertised then for $10 to $25. Bloomingdale's had ladies' dresses, "our own copies of Parisian styles," for $17.50 to $27. Call loans were made at 2 to 2½ percent. *The New York Times* sold for a nickel. Total sales on the New York Stock Exchange on October 25, 1895 were 177,271 shares, and the most active stock—28,250 shares traded—was American Sugar Refining.

It was another world.

And the four men at 35 Liberty Street on that mild and clear October day lost no time in getting to the business of organization. Besides electing officers, they named golf, house, constitution, and by-laws committees, agreed to lease the land and clubhouse from Keller for 15 years, and approved an application to join the United States Golf Association, which had been formed the year before. The club would not be incorporated until 1898.

The clubhouse, a converted farmhouse, was only a few hundred feet from Shunpike Road where Baltusrol Way crossed it. Shunpike had been carved out of the woods and hills in 1801 by locals who wanted to avoid paying the tolls then charged for travel on the Morris & Essex Turnpike—now Morris Avenue.

The first floor of the clubhouse had a large room with pale gray plaster walls and ceiling and white woodwork. It had a stove and a fireplace and was almost entirely enclosed by glass. Meals were served there at small tables. There was also a reception room decorated in red and white, a grill room in green with black woodwork, and a hall in yellow and white. There were rooms upstairs, plenty of lockers, and reportedly adequate bath facilities, although in 1901, when an addition to the clubhouse was being made, Keller was "told" to put a bathtub and a shower in the ladies' dressing room. Electric lights were installed in the clubhouse in 1898.

The nine-hole golf course George Hunter had laid out was not much, even by the standards of that day. Its total length was only 2,372 yards, and only two holes measured more than 300 yards. Hole by hole it went like this: first, 310; second, 260; third, 213; fourth, 233; fifth, 186; sixth, 178; seventh, 187; eighth, 288; ninth, 517. It's possible that the reason for that "monstrous" ninth hole was that Hunter wound up with the eighth green a long way from the clubhouse and had to get back to it with one big hole. Whatever the reason, it was quite out of keeping with the rest of the course. In those days of the gutta percha ball and of clubs you would not have in your bag today, it was a tough finishing hole.

Although the course may have left something wanting, Keller's concept of the need for such a club as Baltusrol was right on the mark. It was instantly popular. The membership went from the original 50 to 137 by December 30, 1895, to 312 by March 1897, and to 400 by February 1898, give or take a few.

Numbers, however, were of minor importance. From the outset, Keller and the board of governors set the norm for the membership, a norm it still enjoys today. They wanted people of substance, of talent, and of accomplishment—and they got them.

Among the early members, in addition to Weeks, DuFais, and Turnure, were many who met that norm. There were Louis P. Bayard, manager of the Phoenix Assurance Company of London, and William B. Beekman, a governor of the New York Stock Exchange and treasurer of the company that made ticker tape. Beekman was married to Keller's sister. There also were Henry Pennington Toler, a broker; William A. Larned, son of a Summit and Newark banker and later to become a seven-time national singles tennis champion. There were J. Walter Wood, a broker and a Keller cousin, and William Fellowes Morgan, a man of many talents and a deep interest in golf. Morgan was an early secretary and later treasurer of the USGA. His wife was the Women's Metropolitan Golf Association's first president (1899–1903), and she was a prominent early member of Morris County Golf Club as well.

Then there were Charles Bassett, vice president of the East River Savings Bank, director of the Franklin Trust, and a director of the Celluloid Company; George C. Kobbe, member of the law firm of Roosevelt and Kobbe and the son of the Counsel General in New York of the Duchy of Nassau; Daniel C. Boissevain, a Dutch American banker and a member of the Calumet Club, where Keller lived, and Carroll P. Bassett, a civil engineer and utility magnate whose family name appears twice in the Magna Carta.

And there were prominent women among the early members: Mrs. John C. Wilmerding, a great-granddaughter of Commodore Vanderbilt, and Mrs. Clement C. Moore, whose husband was a great grandson of the clergyman who wrote "A Visit from St. Nicholas," better known as "The Night Before Christmas."

Actually, the original invitation for membership had included a list of patronesses, and many single women became members of Baltusrol in its early years—for example, Miss Bigelow in 1911, Miss E. Ogden in 1913, and Miss Violet Miller in 1914. Women apparently were more prominent in U.S. golf's early years than they were later. The early popularity of Southampton's Shinnecock Hills Golf Club is

Baltusrol—winter of 1906

17

credited to the influence of the women members "whose enthusiasm knew no bounds." And there were three women on the first executive committee of Montclair Golf Club. Indeed, women founded the Morris County Golf Club in New Jersey in April 1894 and were its first officeholders. And Morris County hosted the second women's golf championship in the United States in October 1896. According to H. B. Martin's *Fifty Years of American Golf*, it was not the purpose of the Morris County women to bar the men, who were permitted to join upon the payment of a suitable fee, it being generally agreed they would make "good caddies."

But the original Morris County women officers were deposed in January 1895 by a slate composed entirely of men—and women's relative importance in golf declined for many years. Only recently have they seriously challenged men's domination of golf's middle years. Some have suggested this started with Babe Didrikson Zaharias' ascendancy in women's golf during the late forties and fifties followed by Mickey Wright's stunning record in the early 1960s. But television has helped: CBS pitted its Sunday afternoon coverage of the 1984 Women's LPGA Tournament directly against NBC's coverage of the Men's Kemper Open, and according to Nielsen ratings, CBS captured 42 percent of the golf-watching audience.

The first golf tournament at Baltusrol took place in February 1896. A cold snap had stiffened the turf into frosty ridges and hollows, but a sprinkling of sand made the greens fairly level. There were forty-five contestants representing other early golf clubs including Philadelphia Country Club, Richmond County Country Club, St. Andrews, Westbrook, Knollwood, Morris County, and Shinnecock Hills, as well as Baltusrol. The winner was W. W. McCawley of Philadelphia. He had a score of 96 gross and 86 net.

Because of Baltusrol's rapidly increasing membership, Keller's original 9 holes were expanded to 18 in the next three years, the enlarged course being ready to play by July 1898, according to the *Commercial Advertiser*. The course was to be rebuilt later—in 1905. The rebuilt course would be known as the Old Course.

Although the board of governors was assiduously devoted to business, a golfer's sense of humor occasionally popped up in official records. The minutes of one 1898 meeting indicated, for example, that "the meeting had been called for 4 o'clock, but as the sun had beamed through the clouds and the course was in fair condition, it was found impossible to assemble until 4:30." Upon convening, the president announced that "owing to the approach of darkness on the links, there was a quorum present."

Willie Anderson

Coming of Age: Keller–Bayard (1898–1919)

Louis P. Bayard, the manager of the Phoenix Assurance Company, succeeded Arthur D. Weeks as president of the club in 1898 and held the post until 1919—a total of 21 years. But Louis Keller as secretary continued his proprietary role in the club and quietly dominated its affairs from his office in New York.

In 1900 Baltusrol joined with five other clubs to found the New Jersey Golf Association. L. E. Graham, a Baltusrol member, served as the association's first president.

The club was just 5 years old in October 1900 when it hosted the largest women's tournament held in this country up to that time. Because of the size of the field, 113 entries, and the prominence of the players, the tournament was ranked on a level with a national championship. The players included Frances C. Griscom of Philadelphia, winner of the 1900 U.S. Women's Amateur Championship at Shinnecock Hills; Beatrix Hoyt of Shinnecock Hills, national champion in 1896, 1897, and 1898; Ruth Underhill, national champion in 1899; Margaret Curtis, who would win the national championships in 1907, 1911, and 1912; and Georgianne Bishop, who would win the Women's Amateur Championship at Merion Cricket Club in 1904. Miss Griscom won the tournament by defeating Miss Curtis in the semifinal and Miss Bishop in the final.

In the following year, 1901, the club was the scene of the U.S. Women's Amateur Championship. It was won by Genevieve Hecker, who would also win it the following year at The Country Club in Brookline, Massachusetts. She was considered one of the best women golfers in the world and wrote the golf primer of the time, *Golf for Women*.

Two years later, in 1903, Baltusrol hosted its first U.S. Open. Willie Anderson won it in a playoff with David Brown. Anderson, who had come from Scotland, had been the club's professional for about a year, in 1898–1899. He also won the U.S. Open in 1901, 1904, and 1905 to establish a record of four victories, which stood for 25 years before it was equaled by Bobby Jones and later by Ben Hogan and Jack Nicklaus.

It is amusing how many "Willies" turned up from Scotland in those early years. Along with *Willie* Anderson, there was *Willie* Park, Jr., who won the British Open in 1887 and 1889 and who played *Willie* Campbell at Brookline in 1895. Then there was *Willie* Dunn, who came to Shinnecock Hills in 1901, and *Willie* Tucker of St. Andrews, *Willie* Smith of Midlothian Golf Club, Chicago, and *Willie* Norton of Lakewood.

In 1904, the U.S. Amateur Championship was played at Baltusrol. It was won by 20-year-old H. Chandler Egan, who would carve out an impressive series of victories, including a successful defense of his U.S. Amateur title in 1905.

Thus began the tradition that has linked Baltusrol so closely to the growth of American golf for almost a century.

The Railroad Connection

Early transportation to and from the club invariably involved horses and carriages—and bicycles of course. In 1896 the board authorized Louis Keller to

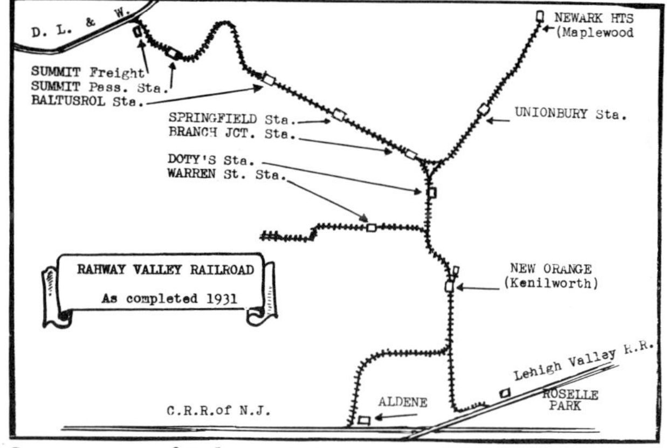

System map of Rahway Valley Railroad

Locomotive of "Baltusrol Special"

Baltusrol Station—Rahway Valley Railroad

provide feed for members' horses at 50 cents per day and to keep two horses at the club for picking up New York members at the Lackawanna Railroad stations in Short Hills and Summit. Both stations were roughly 2 miles from the club.

To shorten this haul, the ever-enterprising Louis Keller purchased the New York and New Orange Railroad in 1904. It had been formed in 1897 but had gotten into financial trouble. Keller reorganized the company, changed its name to Rahway Valley Railroad, and built a passenger station and siding where the track crosses Baltusrol Way about a quarter of a mile east of the club. According to Bernard Cahill, Rahway's chairman, club members coming from New York by ferry could board a "through" car at Jersey City on the New Jersey Central Railroad, connecting with the Rahway Valley Railroad at Aldene and proceeding to the Baltusrol station. Or they could take the Rahway Valley Railroad from Summit to the Baltusrol station. The Summit station was on land now occupied by Overlook Hospital, only a long block from the Lackawanna Railroad's Summit depot.

The 2-locomotive, 10-employee, 12-mile Rahway Valley Railroad was reportedly used by Thomas Edison in making his early films—*The Midnight Flier* and *The Switchman's Daughter*. It still serves indus-trial plants in the Kenilworth–Roselle Park area—Monsanto Chemical, among others. But the Summit branch of the railroad, although still intact, no longer operates regularly. And the Baltusrol station Keller built in the early 1900s burned down after World War II. Many of today's stockholders of Rahway Valley are Keller heirs—descendants of Keller's sister, Mrs. William B. Beekman.

The Night Fire

On the night of March 27, 1909, the clubhouse was destroyed by a fire that had started in the kitchen. It was discovered at 1:00 a.m. by Mary Moriarty, the housekeeper. Mary screamed, waking up John May, the clerk. Together they alerted Jacquot, the steward, who in turn sent a boy to awaken Henry Toler, a broker who was the only member among the 15 people who were sleeping in the building. Toler, reluctant to be awakened from a sound sleep, muttered, "Don't want to be disturbed. Go away." Jack Tennanbaum, the bartender, and John Brush had to get a ladder to get Toler out through a second-story window.

Crossmier Zizi, the chef, had long feared fire and kept a rope ladder in his room. He hung it out the window and started down but either missed his footing or

Leaving Short Hills Station of Lackawanna Railroad

the rope broke and he fell, breaking both ankles. He was the only casualty.

Meanwhile, Toler, once on the ground, reportedly "went back and began dragging out the women who were there, swinging them through windows to the ground." A mighty man was Toler!

The Springfield Hose Company came on foot from a mile away, but the water pressure was so low they could only watch helplessly while the frame building burned to the ground, leaving only the chimney standing. But 12 horses were saved along with paintings, curios, and some trophies.

Toler, the broker who lived at 25 Broadway in New York, was named by the *Herald* as the "hero of the night fire."

A favorite Baltusrol locker room tale has it that one of the members, an errant husband, returned to his New York brownstone home on Sunday, March 28, 1909, with the excuse that he had been delayed in New Jersey and had spent the night at Baltusrol. It is said that his wife responded by handing him the newspaper describing the destruction of the clubhouse the night before. (The reader is warned that this apocryphal story could turn up at any club that has had a fire!)

Promptly after the fire, the fire insurance on the old clubhouse was settled at $60,012, and architect Chester H. Kirk, a member, and contractor Charles L. Bell contracted to build a new clubhouse to be completed in ten months. The new clubhouse, a majestic structure of English manor house design, reportedly cost about $100,000. It was financed in part from voluntary subscriptions by members. The second story of the kitchen wing with the governors' room and business offices was added in 1914. The next major addition to the club, the men's grill, would not appear until 1928.

It is interesting to conjecture what might have happened if the old clubhouse had *not* burned down. Conceivably, the converted farmhouse might still be in use, with endless appendages no doubt.

An automobile shelter had been built on the club property in 1908. It was followed with a 24-car garage building in 1912. For the transition from buggies to cars was sudden—where 90 percent of the members had come to the club by one-horse conveyance as late as 1911, 75 percent came by private automobile as early as 1914.

Distinguished Visitors

Although Baltusrol had many notable visitors through

the years, one stands out because he was President of the United States and had said, "They tell me that Baltusrol is probably the best course in America." The visitor was William Howard Taft. The time was April 1912. Taft was to speak in Newark on April 10 and at New York's Union Club the next night. He played the Baltusrol course on Saturday morning, April 11, with Arthur B. Chandler, a member of the club, and two others. Chandler was a graduate of Yale, class of '78, the year in which Taft matriculated.

According to the *Elizabeth Times*, Taft did not play a great game—but he had a lot of fun. In the midst of the good-natured contest, he sank a particularly difficult putt, "threw his hat in the air, clapped his hands gleefully, and shouted like a boy just let loose from the humdrum of the school room."

In the club's lobby, Taft shook hands with ex-Senator John Kean, New Jersey Governor Tom Kean's great-uncle. He lunched in a private dining room with a small group of friends, including Arthur Chandler and Paul Starett. (Note that the board of governors decided on May 11, 1912, that the club should not have to pay for the President's lunch since he came as a private guest and not on invitation from the club!)

In September 1913, two more distinguished visitors came to Baltusrol: British golfing celebrities Harry Vardon and Ted Ray. (Harold Hilton, another British star, who had won two British Opens and four British Amateurs, had visited the club in 1911, the year he won the U.S. Amateur Championship.) Vardon and Ray played George Low, Baltusrol's professional, and Alex Smith in a special match at the club. Rivalry was still intense between British and American golfers at that time.

Vardon had won the British Open in 1896 and repeated it in 1898 and 1899. He was to win that important event three more times, setting a record that has never been broken—although Tom Watson almost matched it in 1984. Vardon had been making tours in the United States from 1900, playing against the best American golfers and otherwise promoting the sport. He also won the U.S. Open in 1900.

Ted Ray, who accompanied Vardon, was only a shade less skilled than Vardon. Ray won the U.S. Open in 1920 at the age of 43.

Alex Smith, George Low's partner, had come to the United States from Carnoustie, Scotland, in 1898 and was employed as greenkeeper and pro by the Washington Park Club in Chicago, it being customary at that time to assign both functions to the same person. He transferred to Nassau Country Club at Glen Cove, Long Island, in 1901. Smith was second to Fred Herd in the 1898 U.S. Open and second to his close friend Willie Anderson, Baltusrol's former pro, in both the 1901 and 1905 Opens. He won the 1906 and 1910 Opens himself and later became one of the most celebrated teachers of golf—truly a pro's pro.

George Low, who was also from Carnoustie, had followed Alex Smith to America in 1899. He had tied for second place behind Alex's brother, Willie Smith, in the 1899 U.S. Open and had other important finishes. Low was hired by Baltusrol in 1903 as the club's pro, greenkeeper, and clubmaker. He held these positions until 1925. He was also a golf course architect and helped design the Westchester Country Club course. Harry Vardon had previously defeated George Low, 5 and 3, in a match in Miami in 1900. Low is credited with that accurate and oft-repeated observation: "Golf is a humblin' game."

The September 1913, 36-hole match at Baltusrol was won by Vardon and Ray, 7 and 6, before a crowd estimated at 1,000 to 2,000. The winners had a better ball of 67 in the morning round.

Harry Vardon, Ted Ray

Alex Smith, George Low

Through the arches of the veranda

This incidentally was just a week before Francis Ouimet, the 20-year-old amateur from Boston, became an American golfing hero overnight by beating Vardon and Ray in the U.S. Open at Brookline.

Baltusrol's Old Course saw two more national championships: the Women's Amateur in 1911, won by Margaret Curtis, and a second U.S. Open in 1915, won by Jerome D. ("Jerry") Travers of New Jersey's Upper Montclair Country Club. But the course had become overcrowded. And it was no longer a suitable challenge for the rising caliber of play that was emerging.

The New Courses

In October 1916, Louis Keller recommended the building of a second golf course to the board of governors. He offered to sell the club the land that it had been leasing from him plus whatever additional acreage would be needed for the second course. The lease, which had been renewed last in 1909, had 17 years to run. The only land owned by the club at that time was the ten acres immediately surrounding the clubhouse itself.

Keller argued that a lease would be more eco-

Red Cross girls circa 1917

Baltusrol Way creeping over the mountain

nomical to the club than a purchase, but he indicated his willingness to go whichever way the club wished. After serious study, the board opted for purchase rather than lease and this was approved by the membership at a meeting attended by 140 members in November 1917. The club then purchased 318 acres from Keller for $154,000—the 146 acres it had been leasing from him plus 172 acres to the west. It also acquired the right to remove Baltusrol Way, which continued through the course below what is now the first tee of the "Lower", ultimately meandering over Baltusrol Mountain.

Keller, of course, was always in a potentially anomalous position in dealing with the other members of the board of governors in such matters. But the record suggests that he turned square corners in all these dealings.

In retrospect, the club's decision to purchase the land rather than continue to lease it was extremely far-sighted. To be sure, the club incurred a substantial debt. But it started the club on its journey to the independence it so treasures today.

Two completely new 18-hole layouts were promptly designed and built by A. W. Tillinghast, a native Philadelphian, who resided in Harrington Park, New Jersey.

The new courses formed essentially the "Upper" and the "Lower" as one sees them today, although certain changes in each have been made through the years in order to keep them attuned to championship play. Both courses have a par of 72. The Lower measures approximately 7,000 yards from the back tees, and the Upper 6,700. And each has its own individuality.

The Upper has more woods, sharper slopes, and trickier greens. The Lower, where the national championships have mostly been played, is somewhat deceiving at first glance. It appears to offer plenty of room to the player, but the number and placement of hazards and the carry required on approaches make it fully deserving of its reputation as a severe test for the average golfer and the expert alike.

The Lower rates among the "Top 20 of America's 100 Greatest Golf Courses" as compiled biennially by *Golf Digest;* it is twenty-sixth in *Golf Magazine's* list of the world's greatest golf courses; and it is "one of the greatest courses on the face of the earth," according to *The World Atlas of Golf.* And several of the holes on the Lower have been selected at various times for listing among this country's outstanding challenges.

Completed in June 1922, the new courses cost about $140,000.

Coal shortages during World War I required the clubhouse to close for the first three months of 1918. Many members resigned in that year, the total membership going from 743 to 693. Dues were remitted for some 31 members who had entered military service, including the Red Cross and the YMCA. And there was a severe shortage of caddies. Forty of the older ones had entered military service or were working in war plants, and those that remained were decimated by the influenza epidemic.

The Calm Twenties: Sinclair–McKnight (1919–1929)

R obert S. Sinclair succeeded Louis Bayard as club president in 1919 and oversaw the completion of the two new golf courses—a monumental job and, as it turned out, a milestone in the history of American golf.

Louis Keller died on February 16, 1922, at 65, ending a long and remarkable regime at Baltusrol. Although, Keller never held a post higher than secretary during the club's first 27 years, he was its driving force. The club was his idea. He provided the land and the early clubhouse; he oversaw the building of the first golf course; he bought a convenient rail facility; and he even supplied the membership itself—his New York *Social Register* had been a primary source of early Baltusrol members. Even in 1906 almost 50 percent of the 688 members still came from New York.

But many prominent New Jersey names had begun to appear on the club's roster, such as Hartshorn, Frelinghuysen, Auchincloss, and Day, and by 1913 Baltusrol had become a predominantly New Jersey club, with 71 percent of its 718 members coming from the state. And as new golf clubs continued to proliferate in the metropolitan area, Baltusrol's primary constituency ultimately narrowed to the immediate New Jersey communities, principally Short Hills and Summit. Indeed, 96 percent of Baltusrol's current golf members reside in New Jersey. Only 4 percent live out of state.

Baltusrol's headquarters, which had long been located in New York at Keller's office at 29 Broadway and later at 315 Fourth Avenue, were finally transferred to the club itself in 1930 after having sojourned briefly at two other New York locations.

A bronze bust of Louis Keller commemorating his unique role in the founding and development of Baltusrol was sculpted by W. W. Renwick late in 1922, and it stands in the club's lobby today.

Keller's passing left a big hole in the club's management. Although membership on the board of gov-

Robert S. Sinclair

Stewart Hartshorn, Jr.

ernors had shifted periodically during the first 27 years, the club was run largely by Louis Bayard, the 21-year president, with Keller, the 27-year secretary, monitoring most of the policies and handling most of the details—almost as if he were a highly paid professional manager. Keller had to be replaced. This was no job for a part-time volunteer.

The shift toward an on-site professional manage-rial staff began inocuously enough in 1923, the year following Keller's death, when Major R. Avery Jones was hired as greenkeeper.

Major Jones—"Mr. Outside"

Jones had been a major in the British Army in World War I. Reportedly, he had been eq-uerry to the Prince of Wales. A little under medium height with a ruddy complexion and a small mustache, Jones was described as dictatorial and tough—"a boss all the way." He apparently did not like Americans—he thought they were barbarians. But he was impressed with Baltusrol. Having worked at several other clubs, he said around the time of his retirement: "when you leave Baltusrol, where else is there to go?"

Major Jones did not stay as greenkeeper very long. Within six months after he was hired, the board of governors noticed that the club was saving 15 per-cent to 40 percent on the cost of certain supplies sim-ply because Jones refused to accept commissions from supply houses he was dealing with. Here apparently was an intelligent man, a tough man, and a man the board could trust to run the club under its guidance. In May 1925 he was appointed general manager in charge of all departments and employees, subject only to the board of governors and its various commit-tees. He was a strong anchor to windward in the club's operations until his retirement in 1945, after which another major operator emerged, Carl Jehlen, but more about Jehlen later.

William G. McKnight succeeded Robert Sinclair as president in 1926, the year the U.S. Amateur Championship returned to Baltusrol. It was played on the Lower Course. A brilliant field of golfers com-peted in this tournament in which George Von Elm defeated the incomparable Bobby Jones, 2 and 1.

In 1927 the Keller estate needed funds and ac-cepted $94,500 in full payment of the $125,000 bal-ance owed by the club on its 1918 purchase of land from Louis Keller. The club borrowed $100,000 from Prudential Insurance to fund this payment.

The men's grill, a major addition to the club-house, was completed in April 1928 at an approximate cost of $180,000. And the board, astutely recognizing a need to preserve the natural beauty of what had be-come a unique and valuable property, promptly ac-quired an additional 352 acres of land along the Upper Course and over Baltusrol Mountain from the Keller estate for $155,000. In March 1929, the club sold off 178 acres of this land, which were located on the Sum-mit side of the mountain, for $168,845 or $13,845 more than it had paid for the entire 352 acres just four months earlier. The land deals were negotiated by Jo-seph P. Day, a prominent realtor and insurance bro-ker, who was treasurer of the club.

Other, smaller tracts were acquired at that time and later to protect the club against encroachment by real estate developers. One of these was the 1929 pur-chase of the few acres at the rear of the second green on the Lower Course, a good portion of which was, and still is, woodland immediately behind the green and extending along the fifth hole.

Still another was the wooded area at the extreme western end of the club property alongside the sev-enth Upper tee and fairway.

These farsighted additions, together with the ex-panded clubhouse, and the two championship golf courses in their parklike setting, rounded out the ba-sic Baltusrol Golf Club property as it is today. Not un-til the late 1970s and early 1980s would significant refinements be made.

But there was trouble ahead—trouble that would cost Baltusrol 30 percent of its members and over 40 percent of its revenues. Indeed, the club would be hard pressed to survive, let alone meet its debt payments and maintain its property at a decent level. The Great Depression struck after the stock market collapse in October of 1929. And before long the country had numerous bank failures and other business collapses, 25 percent unemployed, and the beginning of President Franklin Roosevelt's New Deal and its alphabetic entitlement programs. Golf clubs, being luxuries rather than necessities, were among the first to suffer. Many folded as a result of the Depression.

Fortunately for Baltusrol, an unusually strong leader took over in 1929—J. Stewart Baker. He was followed in 1938 by another unusually strong leader, Caxton Brown.

Caxton Brown and Stewart Baker

The Depressing Thirties and The Tight Forties: Baker, Brown and Major Jones (1929–1943)

J. Stewart Baker, 1929–1938

J. Stewart Baker traced his origins to Aaron Burr of Revolutionary fame. His great-grandfather was an original stockholder in the Bank of Manhattan Company when it was chartered in 1799. His father was president of that bank at the age of 34. And having started as a runner in the bank after his graduation from Princeton in 1915 — where he was voted handsomest man in the class — Stewart Baker became the bank's president in 1934. He was chairman when it merged with Chase Bank to form Chase Manhattan in 1955.

Baker was a banker all the way. According to *The New York Times*, "His subdued attire and dignified manner marked him as a banker." But Baker, who had joined the club in 1921, was a golf-playing banker; he played to a 7 handicap.

Baker's nine-year term as president of Baltusrol was second in length only to the term of Louis Bayard, although it was a distant second, since Bayard had served for 21 years. But Baker's nine years were Baltusrol's toughest economically, and they called on all of his financial and management abilities and business acumen.

The membership dropped from 650 in 1930 to 450 in 1934. And the club's gross income shrunk from $223,000 to $124,000 in the same period. Coping with this 44 percent drop in revenues while still preserving the club's high standards of maintenance and operation presented a serious challenge.

But as Stewart Baker pointed out in his February 25, 1935, report to the membership, the club had had a sound financial policy from its inception and it was still in satisfactory fiscal condition. In all its years there had been an excess of income over expenditures and the repairs, replacements, and necessary improvements for the large plant as well as the amortization of the club's indebtedness had been paid for out of this excess. Apparently the club never had to assess its members. With the exception of the original small amount of capital furnished by the founders in 1895 and a voluntary subscription when the existing clubhouse was built in 1909, the growth of the club had been financed entirely from income.

With this down-to-earth financial legacy, and Baker's leadership, the club was able to reduce operating expenses by more than 30 percent between 1930 and 1934, including a 40 percent reduction in golf course maintenance. Moreover, it reduced its funded debt from $174,000 in 1930 to $90,000 in 1935.

Baker credited Major Jones for most of this. In his 1935 report to the members, Baker said, "It was only through [Jones's] determination, skill and constant effort that the overhead was brought down without impairing the condition of [the] courses or the service of [the] clubhouse."

Jones had indeed been a tough operator. And he had voluntarily reduced his own compensation during the Depression years. Moreover, during the Depression he had seized the opportunity to purchase 1,000 evergreen trees from a failed nursery for 10 cents apiece and got the WPA to plant them. These are the pines still seen on the courses. He also started the first turf nursery in 1937. Not particularly interested in clubhouse maintenance or improvements, Jones was the single greatest contributor to the beauty of the golf course surroundings. He was "Mr. Outside." In 1936 Martin's *Fifty Years of Golf* stated that "the Baltusrol Club owes much of its prosperity in recent years to Major R. A. Jones, who had charge of the management since 1924."

Jones's accomplishments notwithstanding, it was the financially minded Baker who piloted the club through the Great Depression. The quiet patrician and the former equerry made an interesting team.

The Chronic Membership Problem

But there was still a job to do in restoring the membership from its low of 450 to a sufficient level to sustain the upkeep, replacements, and improvements that the members demanded. It is ironic that Baltusrol, whose golf courses and clubhouse service are unexcelled by any club in the metropolitan area regardless of initiation fees or dues levels, has found itself fighting for membership throughout much of its history. This was not the first time the membership had fallen below a necessary minimum. It had happened during the 1907 financial panic and during World War I. And it would happen again during World War II and in the late 1960s and early 1970s. Yet Baltusrol has always been one of the best golfing "bargains" in the country.

The club tried to meet the membership crisis in the middle 1930s by waiving the regular initiation fee for a while and by permitting up to 50 new members to pay for their proprietary certificates over a three-year period. It also appealed to existing members to help increase membership by interesting their friends to join the club. It even provided them with a road map showing how accessible the club was to New York City residents and how quickly and comfortably they could drive to it over the new express highways.

The membership drive, though, was not strikingly successful. The quota had been reduced from 650 to 500 by March 1939 because of the decreased debt load, but the membership reached only 445. And the membership shortage persisted through World War II.

Johnny Farrell Hired

Johnny Farrell was engaged as club professional in 1934 after a careful search by a board of governors committee headed by William Palen Conway, then president and later chairman of Guaranty Trust. (In the 1950s Conway would contribute greatly to the improvement of the golf courses as chairman of the committee on structural changes.) His search committee reportedly considered Ed Dudley, professional at Augusta National Golf Club, but decided upon Farrell because of his superior golf record and charisma. Farrell had won many national golf championships in the 1920s, including the 1928 USGA Open in which he beat Bobby Jones in a playoff. In the 38 years Farrell remained in his post, he brought great prestige to Baltusrol and earned and received the everlasting affection and admiration of its members. But more about Johnny Farrell later.

A Special Treat by Paul Whiteman

Baltusrol had many glorious dances and parties but none quite as spectacular as the one in 1935 when club

Farrell defeats Bobby Jones — 1928 U.S. Open Playoff

member Paul Whiteman presented his full orchestra in concert on the lawn. Among the artists who performed that warm July evening were Helen Jepson, Johnny Hauser, Ramona, and the King's Men. Whiteman, who had been nationally recognized for his symphonic approach to dance music as early as 1918, and who had premiered George Gershwin's stunning "Rhapsody in Blue" in Aeolean Hall in 1924, was still a top celebrity in 1935. Indeed, he and his band would star on the *Goodyear Review* and other network television shows as late as the middle 1950s.

After the concert, the Whiteman orchestra played for dancing in the men's grill, probably closing with "When Day Is Done," using a Henry Busse-type muted trumpet. Whiteman's fellow club members expressed their appreciation for his generous musical treat by holding a stag dinner in his honor the following December at which other leading stars of stage and radio were also present.

Sunday night movies on the lawn or the terrace were also very popular during the middle 1930s and the 1940s. Among the many films shown were *The Petrified Forest, Mr. Deeds Goes to Town, The Joe Louis–Max Schmeling Fight* (1936), *The Go-Getter* and *Tovarich* both starring Anita Louise, and *Mad About Music* with Deanna Durbin. Also, shown were *The Awful Truth, Ebb Tide, Her Jungle Love, Wells Fargo,* and *The Rage of Paris* with Danielle Darrieux and Douglas Fairbanks, Jr.

The last token of the high-society aura that Louis Keller had vested the club with at its inception was finally expunged in 1935. The board of governors changed the club address from Short Hills to Springfield, although a reference to the Short Hills station of the DL&W Railroad was retained on the letterhead. One can only surmise that Keller must have felt that Short Hills was a "better" address socially than Springfield. Perhaps it was. But in the intervening years, golf had gone from being solely a rich man's game to being the favorite sport of millions of Americans. This change notwithstanding, there seem to be as many if not more captains of industry and finance among Baltusrol's members today. It is the social label that has faded.

But back to golf.

The U.S. Open Championship was played at Baltusrol in 1936. J. R. Monroe, manufacturer of the calculating machine, was general chairman of the event. It was the only USGA event held on the Upper Course, [although the 1985 U.S. Women's Open will have been played there.] Tony Manero, a near-unknown until that time, won the 1936 tournament. Incidentally, it was Ben Hogan's first National Open, although he missed the cut.

The 1936 Open was planned and conducted along the same lines as Opens today: Members were expected to support it by buying admission tickets, access to the clubhouse was limited, car parking was tightly controlled, and so forth. The Open reportedly netted the club $17,000.

The solarium in the men's locker room was extended outward right after the 1936 Open, with most of the work being done by green personnel.

Stewart Baker stepped down from the club presidency in October 1938 and was shortly thereafter honored at a club dinner for which the board of governors waived the ban it imposed in 1933 against female entertainers.

Baker died in 1966, but his wife, Marianne, is at the time of this writing, still an ardent golfer and one of but five current honorary members of the club. The others are Johnny Farrell, Carl Jehlen, Jack Nicklaus, and Mrs. Maurice N. Trainer.

Caxton Brown   1938–1943

Except for the precious few years of "normalcy" in the late 1930s, the entire Stewart Baker–Caxton Brown era was plagued by major problems — Depression-induced in Baker's case and World War II shortages in Brown's.

But Caxton Brown was a man for his time too. A graduate of Columbia, Brown joined Weston Electrical Instrument in 1901 and was its president from 1944 to 1947, when he became chairman of its executive committee. His office was in Newark and, according to octogenarian member Henry Hoyt, Brown faithfully dropped in to the club on his way home from work every afternoon to check that everything was all right. And everything *was* all right in the first three years of his presidency. The high spot was the highly publicized dinner and entertainment on April 12, 1941, to celebrate the final payment and extinction of the club's debt that traced back to the purchase of the land from Louis Keller in 1918. Around 200 members assembled to witness the burning of the mortgage.

But everything was *not* all right in Caxton Brown's fourth year. The effects of World War II were felt almost immediately as the nation mobilized to battle Japan and Germany. The number of golf members dropped to 429 by January 1942 (403 by April 1943), and dues were remitted for those in the military services. The golf courses and club services deteriorated as shortages developed in manpower, gasoline, food, and liquor. Even golf balls were in short supply because of the cutoff of rubber imports from Malaya. The USGA urged members to turn in old balls for reconditioning. And Baltusrol's tournament committee changed the drop rule to prevent loss of more than one ball on the fourth Lower.

But despite the need to curtail maintenance, the golf courses were kept in play. And five horse-drawn conveyances and four teams were purchased to transport members between the club and the Short Hills railroad station. More than 140 members were transported on some Saturdays and Sundays. The two largest vehicles each had a capacity for 24 persons. The last of the four pairs of horses was sold in 1944.

An animal husbandry program was installed in 1943 with the blessing of the U.S. Department of Agriculture. And Victory gardens were planted. Unused land between the fifth, thirteenth, and fourteenth Upper fairways was prepared to raise hay, with 8 acres being devoted to crops and 1½ acres planted in vegetables. In the spring of 1943, the club acquired some 30 well-bred steers, 13 Guernsey cows with calf, and approximately 80 sheep. The sheep were helpful in "mowing" the grass on the fairways, but their sharp little hooves cut up the greens, paticularly on the fifth Upper. The club sold the livestock in the fall of 1943 at a mild profit, but this was deemed far less important than the contribution that had been made to the national food supply and the highly publicized example that was set for other clubs and institutions to follow. Moreover, the farming activity qualified the club for additional rations of hard-to-get gasoline.

Major Jones again was the driving force in imaginatively meeting all of these World War II problems.

The club sold 51 acres of surplus acreage on the other side of Baltusrol Mountain to Diamond Hill Corporation in March 1941, and it deeded a 60-foot strip along the top to Springfield Township to build Montview Road. But it still owned the 18-acre Sweeney Farm across Shunpike opposite the 1st fairway Lower, which it had bought in 1937 to prevent an unsightly real estate development there. Otherwise, the club's land acquisitions and sales had pretty well stabilized.

Caxton Brown was succeeded by Walter R. Hine as club president in November 1943. Brown died nine years later in 1952, but at the time of this writing, his daughter Elaine, the wife of a prominent member, Harvey E. Mole, was still active in the club.

Burning the mortgage, April 12, 1941

Restoration: The Steady Ones and Carl Jehlen 1943–1958

Walter R. Hine   1943–1945

Walter Hine lived in Short Hills. He was a vice president of J. Walter Thompson, the big advertising agency. His principal account was Eastman Kodak.

Hine's term as president was uneventful except that the war ended and the club celebrated its fiftieth year with an anniversary dinner in 1945 — and Major Jones resigned in July of that year, apparently after a disagreement with Hine. Jones was succeeded as club manager by Tom Burns, a graduate of Cornell University's Hotel School, and as green superintendent by Ed Casey, who remained until late 1967. But in November 1945, after only two years in office, Hine resigned as president and left the club as well, reportedly because of the bitter criticism that had been leveled at him by the vociferous admirers of Major Jones.

In truth, Jones probably had outlived his usefulness to the club. He had been a tower of strength during the Depression and World War II. His role in controlling club costs while still preserving the basic golf courses and enhancing their surroundings was immense. But in the process the structures had become severely run down. The garage and the employees' quarters above were a shambles. As many as seven men lived in a single cramped room. The clubhouse, so highly touted only 15 years before, had deteriorated from lack of maintenance. Paint had peeled, wood trim had rotted, and pipes had rusted.

Despite his undoubted strength and talents as Mr. Outside, Jones was not the person to head up the structural rehabilitation that was sorely needed or to manage service personnel in the clubhouse. It would take someone with considerably different abilities to manage the post-Depression, postwar recovery and to build up the club again, although that person would not appear for another eight years.

East Lounge — 1941
Foyer — 1941
Mens Locker Room Solarium — 1941
Dining Room — 1941

Maurice N. Trainer   1945–1947

Hine was succeeded as president by Maurice N. Trainer in November 1945. Trainer was born in East Orange in 1888 and was a graduate of Penn Charter in Philadelphia and the University of Pennsylvania with a degree in electrical engineering. He was president of American Brake Shoe Company. His son, Bob, was a member of the club at the time of this writing, having joined in 1959. And his widow, Roberta, who when last contacted was in her nineties and living in Summit, was made an honorary member.

In a letter to members in January 1946, Maurice Trainer catalogued the problems and laid down a course of action to restore the run-down golf courses and clubhouse to their former condition. Rebuilding the membership and, particularly, attracting new house members were high on the agenda, and good food and good restaurant services were cited as the key to this.

Progress was made in the following year. The membership increased by 48 or almost 10 percent; 113 servicemen, mostly sons of members, had returned; and progress was made in restoring the golf courses through increased green personnel and the purchase of new equipment. Indeed, both the fairways and the greens were expected to be the equal of any in the area by 1948. Also, kitchen equipment was replaced, and the kitchen floor was resurfaced to improve cleanliness. The small lounge was redecorated and new slipcovers placed on the furniture; the ladies' powder room was enlarged and redecorated. The roof over the main dining room was waterproofed. And major repairs and improvements to the heating, plumbing, and electrical systems were made. All pretty mundane stuff, but then 15 years of under maintenance had taken a heavy toll!

Meanwhile, Shepard Barnes was named chairman of the 1946 U.S. Amateur, which was played on the Lower Course in September. Barnes was also president of the Metropolitan Golf Association in 1948–1949. Ted Bishop defeated Smiley Quick in that tournament in a down-to-the-wire final round. The net profit for the club for the Amateur was reported to be $5,536.

Because of the demands of his job as president of American Brake Shoe Company, Maurice Trainer resigned as president of the club in October 1947, after just two years in office. He died in 1968 at age 80.

Stoddard M. Stevens   1947–1952

Trainer was succeeded as president by Stoddard Stevens, who had been a partner of Sullivan & Cromwell since 1929. Stevens was born in 1891 and graduated from Cornell University in 1914 and from Columbia Law School in 1917. He was a highly respected professional.

The Stevens administration methodically continued the extensive renovation program begun by Maurice Trainer. The ladies' locker room, which had earlier been moved to its current location across from the main entrance of the club, was extensively renovated in 1948. Additional motorized equipment was purchased for the golf courses and cyclone fencing and traveling sprinklers were installed. The parking lot was paved with hardtop.

Also, structural changes in the golf courses recommended by Francis Ouimet and Robert Trent Jones were made in 1949 and 1950 by Palen Conway's committee on structural improvements. The purpose: "To make the courses fairer for the average player and harder for the low handicapper." And the practice green was enlarged to 18 holes.

In 1951, at the suggestion of Philip Hartung, chairman of the golf committee, practice and teaching on the first fairway of the Lower were curtailed and a new practice range was opened behind the garages. Another practice range was set up in 1953 on land owned across Shunpike Road, but it was soon abandoned because it was practically never used. A blinker light was installed at the busy Shunpike entrance about the same time. Hartung, a 1923 Lehigh graduate, was a vice president of Public Service Electric & Gas Corporation.

Stoddard Stevens reported spending a total of $200,000 for renovation and improvements during the five hectic years from 1947 to 1952. Stevens was 90 when he died in Short Hills in 1981.

But there was much more to be done before the golf courses, the physical structures, and the members' services would attain the degree of perfection that is so evident today. And every succeeding administration continued in its own way to make some contribution toward this goal—renewing, modernizing, strengthening, improving, and otherwise enhancing the club's beauty and reputation—almost as a succession of devoted teachers might enhance the abilities and knowledge of a talented youth. The Ramsey ad-

Ed Sullivan Show—Johnny Farrell, Collette Ramsey, Ed Sullivan, Kay Farrell, Hobart Ramsey—1954

ministration, which took over from Stoddard Stevens in 1952, was no exception.

Hobart C. Ramsey   1952–1955

Hobart Ramsey was born in 1891. He graduated from the U.S. Naval Academy in 1915 and was a lieutenant commander in the battleship and destroyer forces during World War I. He was a member of the War Manpower Commission during World War II.

Chairman of Worthington Pump and a founder of the Navy Industrial Association, Ramsey was also a director of Prudential Insurance Company, Armstrong Cork, and New Jersey Bell Telephone. He was given the Distinguished Service Award by President Eisenhower. Ramsey, who was an intercollegiate wrestling champion, played golf to a 6 handicap. And he was married to one of the best woman golfers in Baltusrol and indeed the whole metropolitan area— Collette Ramsey, who donated the Ramsey Tournament Trophy and who at the time of this writing was still living in the Short Hills area.

During the Ramsey period, the parking lot was enlarged and a small exit provided at the lower end so that cars would not have to go around the main entrance of the club. It was promptly dubbed "Ramsey's Alley." A watering system was installed on the Upper Course, and a shed was built for machinery. Also, a new and larger bar was constructed in the men's locker room in 1955, with photographic murals behind it. And the main dining room was air-conditioned at a very favorable cost to the club, Ramsey having been engaged in the manufacture of air-conditioning equipment.

Also during the Ramsey years, the 1954 U.S. Open was played on the Lower Course after a significant facelifting by Robert Trent Jones. C. P. Burgess was chairman of that event.

This tournament proved to be one of the most significant events in the history of the USGA. It ushered in a new era of spectator golf. The attendance was, of course, a record at that stage, and the land owned on the other side of Shunpike was devoted to car parking. Roped fairways controlled the surging gallery, a problem that was increasingly plaguing big golfing events. (Ropes were reportedly used to control crowds as early as the 1898 U.S. Amateur, which was held at Morris County Golf Club.) More dramatic, however, was the fact that the 1954 Open was the first such tournament to be shown on national television, and a new pattern of spectator interest from all walks of life sprung forth. The unique character of this tournament was fittingly climaxed when Ed Furgol, a golfer with a withered left arm, won the championship by playing the eighteenth hole of the final round in an unorthodox but effective manner. Completely stymied in the woods after his drive, Furgol found daylight to the eighteenth fairway on the Upper Course and traveled that route to get his par and clinch the title. The club's gain from the Open was reported to be $78,000.

Hobart Ramsey died in his nineties in Short Hills.

The Ramsey administration's most epochal action, though it was little heralded at the time, was the employment on March 1, 1953, of Carl J. Jehlen as club manager.

Carl J. Jehlen—"Mr. Inside"

Jehlen was born in Germany in 1906 and came to the United States in 1909. He was educated in Paterson, New Jersey, and New York City schools. Before coming to Baltusrol, he had been engaged in hotel and club work for 31 years, starting as an assistant bookkeeper in the Biltmore Hotel in 1922 and working his way up to the managership of the prestigious Glen Ridge Country Club in New Jersey. He was president of the Club Managers Association of America in 1953 and considered one of the top people in his field—a thoroughly financially conscious club manager.

It should be noted that despite the considerable managerial abilities of the string of very important executives who headed Baltusrol, there probably was not a single one among them who had had any meaningful experience in the fine art of club management, and especially in food and beverage service, except, perhaps a stint in the kitchen of a college fraternity house.

Moreover, it had long been accepted in the club world that food services were doomed to be unprofitable and that the profit on beverage sales had to be looked to to offset the food loss. (At one time, of

Carl Jehlen

course, slot machines were also an important source of club revenues, but alcoholic beverage and antigambling laws eventually diminished them.)

To their credit, the post–World War II presidents of Baltusrol realized the need for an expert manager in this highly specialized area. Tom Burns had not quite measured up. Perhaps Jehlen would.

Jehlen lost no time in improving club services. He appointed Ernie Mudd as the new headwaiter in May 1953. And he took a firm hold of the food operation, which had been producing small profits or losses up to that time. Jehlen had it earning $30,000 a year by the 1960s, $40,000 by the 1970s and $150,000 by 1982. He had the benefit of a skilled French chef, Jean Verdi, who was employed in 1955, and in 1985 was still running a fine kitchen. Beverage profits, of course, increased too—but the food operation had meanwhile become an equal profit contributor. Moreover, the Baltusrol dining room's reputation for good food and elegant service expanded to such a point that it has frequently been described as the best restaurant in northern New Jersey.

Jehlen's value to the club was acknowledged immediately by the board of governors. He was invited to sit in on board meetings, and his highly cogent opinions and recommendations were accorded great respect. And he was highly respected by staff and club members as well. In July 1962 Jehlen was given complete charge of everything about the club except the grounds, which were the responsibility of grounds superintendent Ed Casey.

Jehlen was named Club Manager of the Year in 1972 by the regional Club Managers Association and an honorary member of Baltusrol in 1978.

Monroe J. Rathbone   1955–1958

"My one final act, which is worth recording," wrote Hobart Ramsey in 1955, "is that with the aid of several of my associates, I persuaded Jack Rathbone, then president of Standard Oil Company of New Jersey, to succeed me as president so that we might continue the kind of business management we had going."

That *was* quite a feat—to persuade the eventual chairman of the largest industrial corporation in the world to take on the management of a golf club, however prestigious it was, particularly since the job called for meticulous attention to minor detail—something that does not come easily to those accustomed to making world-shaking decisions.

Born in 1900, Rathbone became one of the most important figures in the history of the oil industry. He graduated from Lehigh University in 1921 as a chemical engineer and spent his entire working life of 44 years with Standard Oil. He was also a member of the board of directors of AT&T, Morgan Guaranty, and Prudential Insurance, among others. But his accomplishments in making Exxon truly international were immense. All this notwithstanding, Jack Rathbone, a man of 6 feet, 3 inches, with the battered look of a retired boxer, was very popular at Baltusrol and accessible to all members.

In 1958 the Rathbone administration sold for $90,000 the 18-acre Sweeney Farm, across Shunpike which had been purchased in 1937 and which had been used for parking during the 1954 U.S. Open. Otherwise, the Rathbone administration continued along the same lines as its immediate predecessors—stressing the restoration of structures rather than the improvement of them.

But the 55-year-old clubhouse was becoming aged and tired and the members increasingly set in their ways and resistant to change and innovation. In 1957 a Muzak system with 11 speakers was installed in the dining room and turned on—a small innovation. "What's that, music at Baltusrol?" roared a long-time member, hearing it for the first time. "Get it out!" he growled. And the music was promptly turned off, not to be heard again until 15 years later when restaurant patronage was falling off and a new member of the board, a one-time dance musician, suggested that music might help warm up the dining room—"make it more inviting," he said. Music was cautiously reinstated and restaurant receipts rose—which may have been a mystery to some but not to the former musician.

Improvement: The Venturesome Quintet (1958–1973)

William M. Walther

B etween 1958 and 1973, five administrations continued to enhance the club's already considerable golf reputation by spending large sums on the golf courses and hosting two more USGA championships.

Indeed, the most successful Women's Open up to that time was held at Baltusrol in 1961, again on the Lower Course. The success of this event virtually removed women's golf from dependence on the men's support and established it as a full-fledged, self-sustaining activity. Mickey Wright won this one going away, and many of the members and other spectators watched in awe as Mickey repeatedly hit 5-irons solidly to the pin on the par-3 twelfth hole, which played 180 yards for this event—all carry.

In 1963 a spectacular fire fanned by high winds swept the woods from in back of the practice range all the way to the third hole Upper, threatening both the garage and the clubhouse. Large crowds who came miles to see it had to be controlled by the Springfield Police—and the fire itself was brought under control by the Springfield Fire Department, which had significantly upgraded its equipment in the half-century since the 1909 clubhouse fire.

In 1965 the Internal Revenue Service ruled that a club's basic tax exemption would be endangered if its income from outside sources exceeded 5 percent of its total income. Because of the limitations on outside earnings imposed by this ruling and the increasing real estate taxes and other operating costs, president William A. Walther urged members to step up their use of the club.

Robert Finney was chairman of the 1967 U.S. Open, which was played on Baltusrol's Lower Course. Jack Nicklaus won it with a record-breaking score of 275—just weeks after he had missed the cut in the Masters. Jack described his play at Baltusrol as "just about my finest ever." Arnold Palmer, Nicklaus's

Robert Finney—"Mr. Baltusrol"

playing partner on the final day, was runner-up. It was in this tournament that Lee Trevino, who finished in fifth place, first came into prominence in the golfing world. And it was the last Open for the immortal Ben Hogan, who still drove the ball beautifully but who was no longer the Hogan of legend with his irons and was at best a fair putter. Nevertheless, Hogan finished a creditable thirty-second at age 54. The club's gain from the Open was reported at $219,000.

Along with holding the two national tournaments and improving the golf courses during the years 1958–1973, the formerly economy-minded board of governors became increasingly venturesome in making additions and generally dressing up the club itself.

The presidents during that period were:

— John C. Smaltz (1958–1961), International Flavors & Fragrances executive, who joined the club in 1920 and was living in Florida at the time of this writing.

— Walther H. Feldmann (1961–1964), successor of Hobart Ramsey as chairman of Worthington Pump and a top officer of the National Association of Manufacturers.

— William M. Walther (1964–1967), stockbroker and retired member of the New York Stock Exchange, who was born in Brooklyn in 1895 and served in the army in World War I and the navy in World War II, and who became president of the New Jersey Seniors Golf Association as well as chairman of the New Jersey Golf Association Caddies Scholarship Foundation, which was formed in 1947.

— Robert Finney (1967–1970), "Mr. Baltusrol," who played the Old Course as early as 1915 and joined the club upon graduating from Yale in 1923, who had served on various committees of the board and the board itself for over 30 years, and who was a club champion and five-time seniors champion.

— Matthew J. Glennon (1970–1973), a prominent lawyer from Westfield.

During these five administrations, the pro shop was enlarged, the terrace lounge was enclosed, the first- and second-floor bedrooms were air-conditioned (1960), a new parking lot and service road were added behind the garage (1960), the men's locker room was completely refurbished, the dining room was extensively renovated and a new service bar was built (1962), and the cocktail lounge and the women's locker room were refurbished (1964–1965). The kitchen was extensively renovated at a cost of more than $140,000 (1965–1966), a new heating system was acquired (1967), and an automatic irrigation system was installed on both courses for $400,000—this was Bob Finney's pet project. Also, the Farrell Room was outfitted (1970), the east lounge was refurbished (1972), and new carpet was laid in the men's locker room (1973).

The value of proprietary certificates was increased from $500 to $1,500 during this period to help finance improvements, and dues were regularly raised to meet the rapidly accelerating costs of operating a private golf club in a changing social environment. Club manager Carl Jehlen was particularly attuned to the trends and their implications for the future of private clubs. But John S. Roberts was the first club president to aggressively seek a long-term solution.

Expansion: The Merchandisers (1973–1985)

John S. Roberts became president of Baltusrol in 1973. He was president of F. W. Woolworth Company, as well as a courtly gentleman and a fine golfer. But most of all, Roberts was a skilled merchandiser whose approach to a cost-revenue squeeze was not to economize but rather to boldly expand revenue-producing operations. His was the first of four increasingly expansive administrations.

"There is no way we can support the costs of a private club for taxes, services, and other expenses without increasing revenues," he said. "Unless we can develop more activities to increase member usage, it is questionable whether we can maintain our standards or even survive financially." Of course, enhancing the attractiveness of *existing* facilities to induce greater usage was a part of the Roberts philosophy.

The club's cost-revenue squeeze was aggravated by the drop in golf membership from 497 in January 1974 to 468 in March owing to the Middle East energy crisis and the economic downturn that followed. Roberts wryly observed: "Obviously, we have a serious membership deficiency."

Roberts began immediately to explore paddle tennis as a possible additional income-producing activity. But his proposal to build paddle courts was voted down by the board more than a year later. Baltusrol was to remain exclusively a golf club.

Meanwhile, interest developed in enlarging the pro shop. Among other things, this would help attract a name professional.

The first estimate of the investment required was $90,000, but this was increased to $150,000 as the concept expanded. And the enlarged pro shop was finally brought in at more than $200,000. It is one of the largest and most beautiful pro shops in all of New Jersey, and the members are proud of it.

The Roberts merchandising approach to spur-ring member interest continued under president B. P. Russell (1976–1979). "Bobbie" Russell was chairman of Crum & Foster. At the time of this writing Russell, a fine executive and an excellent golfer, is chairman of the USGA Foundation and national chairman of the USGA's Capital Campaign for $10 million, part of which is to be used to build a new headquarters building in Far Hills, New Jersey.

During the Russell regime, Bob Ross was appointed club professional after an extensive search by a board committee headed by John Farrell, Jr. Ross, an attractive and capable person, became still another refreshing influence on the Baltusrol scene.

As part of the golf course upgrading for the 1980 U.S. Open, Russell recommended the building of five rain shelters on the golf courses at a cost of about $10,000 apiece. He offered to donate the first one. His example attracted donors for the other four shelters: Robert J. Boutillier, then vice president of the club; Charles Simpson, chairman of the 1980 U.S. Open; Arthur Snyder; and Mrs. Frederick J. Leuders, in memory of her husband. These stunning fieldstone structures were completed near the end of 1979 and even further uplifted members' spirits.

The kinds of innovations begun during the Roberts' regime and continued under Bobbie Russell's administration received additional impetus under Robert J. Boutillier, who became president of the club in 1979. Boutillier was vice chairman and partner-in-charge of U.S. operations of the Big Eight accounting firm of Peat Marwick Mitchell & Company.

Of course, his administration continued such bread-and-butter projects as again refurbishing the women's locker room. But the board became even more imaginative and expansive as time went on.

One idea, the "Hill Property Development Project," involved the construction of condominiums on land owned by the club at the crest of the mountain. These would be offered for sale to Baltusrol members. Although the IRS had ruled that the project would not impair the club's basic exemption, the project was dropped in 1980 for other reasons.

Meanwhile, the club hosted the 1980 U.S. Open, which was again played on the Lower Course. The chairman was Chuck Simpson. That tournament was won by Jack Nicklaus after a grueling four-day head-to-head duel with Isao Aoki of Japan. Nicklaus's score: a record 272. It was Nicklaus's fourth Open victory—matching the record co-held by Ben Hogan, Bobby Jones, and Willie Anderson. The club netted $22,000 after substantial improvements to the property were made.

Club manager, Carl Jehlen retired on January 31, 1981, with a large reception being held to honor him and his wife, Marie. At that time, it was said: "Only the Governors and especially the ten club presidents he served would know how much untiring effort Carl devoted to his job to maintain Baltusrol's standards and to assist them in making the correct decisions. Managing the clubhouse with all its ramifications—the staff, the food, the beverages, the utilities, the parties, the meetings—seven days a week is a very demanding job. But besides all these details, Carl was ready to receive the members and their guests as host of Baltusrol—every day."

"Mr. Inside" was succeeded as club manager in 1981 by Mark De Noble, who had been assistant manager of the club since 1979.

Mark graduated from Fairleigh Dickinson University in business administration and had been club manager at Suburban Golf Club and Essex Fells Country Club before coming to Baltusrol. He promptly took the same firm hold of Baltusrol's operations as his predecessor and at the date of this writing looked very likely to become Mr. Inside II.

The Clubhouse Project

Proposed plans for major renovations to the club were revealed to the membership by president Boutillier at a meeting on September 19, 1981, attended by more than 200 interested members. The project included a new informal mixed grill for men and women diners, enclosing the terrace, and expanding kitchen and dining facilities as well as the cocktail lounge, the women's locker room, and the solarium. The cost was estimated at $2 million and would be financed in part by the sale of a 200-foot strip of land, about 30 acres, along the top of the mountain.

Because of the negative response to the proposal by those at the membership meeting, the board retreated to a more modest program for interior refurbishing and redecorating, to be financed by raising the value of proprietary certificates from $1,500 to $3,000 and issuing new nonvoting proprietary certificates to House members at $750—which the members agreed to at the 1982 annual meeting. The total

proceeds from the additional financing were about $1 million.

Work proceeded on the revised clubhouse project under Paul J. Hanna, who became the twentieth club president in October 1982. Chairman of Government Employees Financial Corporation and a director of several companies at the time of this writing, Hanna was formerly vice chairman of Government Employees Insurance Company and Geico Corporation. He was a lieutenant colonel in the 101st Airborne Division in World War II, and he was executive vice president of Manufacturers Hanover Corporation. He was the 1984 Augie Kammer putting champion.

In early 1985, work had been substantially completed on both the first floor and the men's solarium and locker room floor, including the new outside patio for mixed dining. But as in the case of the pro shop several years earlier, the project was expanded several times, and unusual deterioration was encountered in the club's plumbing, electrical and heating systems, so that the total cost will exceed $1.6 million.

But with this sweeping renovation, the clubhouse has finally caught up with the pro shop and the golf course. It is beautiful and should greatly enhance members' usage.

Paul Hanna (left) and B.P. Russell (right) with Governor Thomas Kean

The Anchors to Windward

Mark DeNoble

altusrol has had much going for it in its 90 years:

• A far-sighted founder, Louis Keller, who supplied the site, the first golf course and clubhouse, and the membership together with a railroad to transport members to the club.

• A capable and ever-dedicated board of governors, who supplied the guidance and authority needed to keep the club on course and advance it.

• Major Jones and Carl Jehlen, Mr. Outside and Mr. Inside, who managed it, and their successors Ed Casey and Joseph Flaherty, green superintendents, and Mark De Noble, club manager.

• The celebrated Johnny Farrell.

In addition, there were the anchors to windward, an unusually loyal and efficient staff who served members and guests well from the beginning, quickly calling them by name however long since their last visit.

Here are some of the more visible "anchors" who were on deck in 1985:

William Eldridge, cocktail waiter (1946)
Ernest Mudd, maître d' (1949)
Frank Durnien, doorman (1953)
Jean Verdi, chef (1955)
George Baker, caddiemaster (1967)

And finally, Jean Storey, secretary to the club manager, who came to the club in 1968. Jean graduated from Mary Baldwin College in Staunton, Virginia, and Katherine Gibbs in New York. In 1976 she managed the club for six weeks during manager Carl Jehlen's illness and was profoundly thanked by an appreciative board of governors.

William Eldridge

George Baker

Ernest Mudd

Frank Durnien

THE GOLF COURSES

Baltusrol's initial golf course was laid out in 1895. It had 9 holes averaging 250 yards with 40-foot greens — according to Louis Keller, who founded Baltusrol. The original 9 were expanded to 18 holes in the next three years and were ready to be played by July 1898, according to the *Commercial Advertiser* newspaper.

The expanded original course was the scene of the Women's National Amateur Championship in 1901, which was won by Genevieve Hecker, the National Open in 1903, won by Willie Anderson, and the U.S. Men's Amateur in 1904, won by H. Chandler Egan.

In 1905 the expanded original course was superseded by what is now called simply the Old Course.

The "Old Course"

The Old Course was bounded on the east by Shunpike Road; on the south by the fence line of today's second hole Lower, thence to the left side of the current fifth fairway, and continuing along the row of trees at the left of the fifth green all the way to Hillside Avenue (then known as Turkey Lane); on the west by Hillside Avenue straight up the hill past what are now the greenkeeper's office and storage sheds and to the spring between the fourth green and the fifth tee Upper; and on the north by Baltusrol Mountain, of course.

Baltusrol Way still meandered through the property — in front of the golf shop, across fairways, and finally up and over Baltusrol Mountain. It is said that it was far more difficult to get a ball out of this old dirt road (no free lifts were allowed) than out of a bunker because of the deep ruts left by wagons plying back and forth.

The Old Course was only 6,189 yards long. But according to Bob Finney, a former president of the club, it did not play easily, although it was not as difficult as either of the current courses. There were not too many traps or bunkers. Each hole had a descriptive name. Finney and Fritz Leuders, another prominent member, furnished a description of the course in 1965 for the *Baltusrol News*. The captions under the accompanying pictures were based on that description.

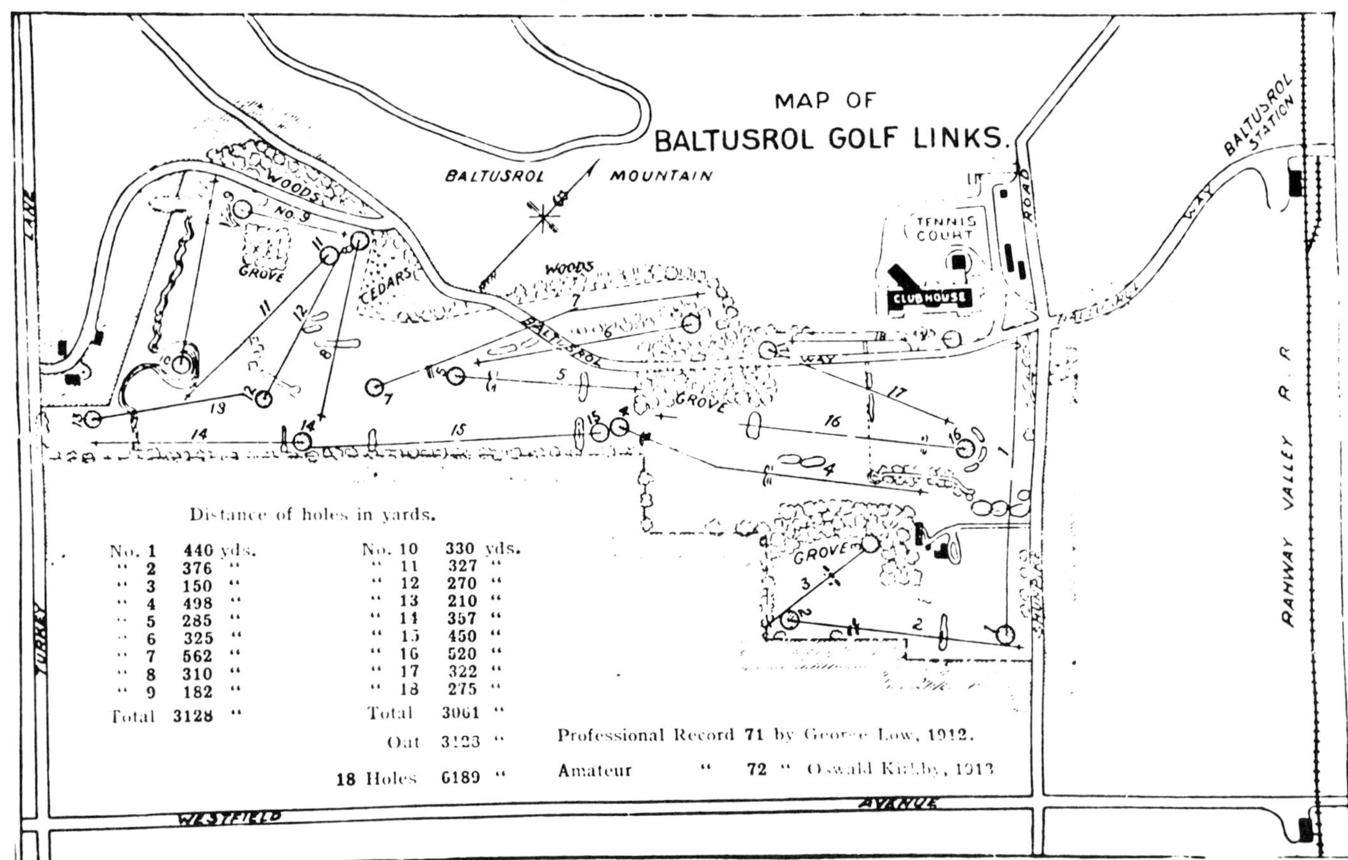

1913 map of old course

The Old Course saw the U.S. Women's Amateur in 1911, which was won by Margaret Curtis, as well as a second U.S. Open in 1915, won by Jerry Travers.

The course record of 71 was set in 1912 by George Low, Baltusrol's resident professional. Five players, including Chick Evans, tied that record during the 1915 U.S. Open, but no one ever broke it.

Starter's shanty

Caddy house and work shop

45

The 1st hole, 440 yards, par-5, was essentially the same as the present 1st Lower.

1st tee

1st green

The 2nd hole, 376 yards, par-4, was essentially the same as the present 2nd Lower.

2nd tee

2nd green

3rd fairway

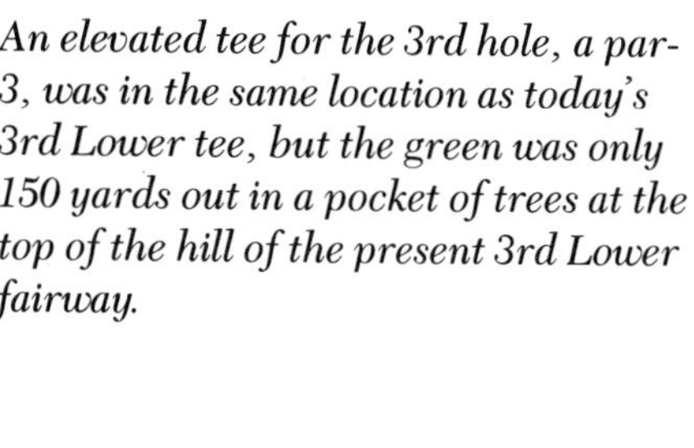

An elevated tee for the 3rd hole, a par-3, was in the same location as today's 3rd Lower tee, but the green was only 150 yards out in a pocket of trees at the top of the hill of the present 3rd Lower fairway.

3rd green

4th tee

4th fairway

4th green

5th tee

6th tee

6th green

7th steps

7th tee

7th green

The longest hole, the 7th, stretched 562 yards—a fine par-5. It started in the woods behind the current 2nd tee Upper, and the drive called for a shot to the left side of the current 2nd fairway. The green for this 7th hole, a slight dogleg to the left, was about where the women's tee is for the current 17th Lower.

The 8th hole was an interesting but short 310 yard, par-4. The tee shot was in the neighborhood of the current 15th tee Lower, toward the mountain, with the second shot going to a very elevated green, completely bunkered in front, which was located behind the current 3rd tee Upper.

8th tee

8th green

Still further up the mountain was the 9th tee. The 9th hole, a 182-yard, par-3 was, except for the location of the tee, about the same hole as the current 3rd Upper.

9th tee

9th green

*The 10th hole was known variously as
the "Pond" hole and the "Island" hole.
The tee was fairly well up into the woods
behind the swamp, at the rear of the
current 3rd green Upper and at the
right of the current 4th fairway Upper.
A swamp, full of water, was immediately
in front of the tee. The green, now the
16th Lower, 330 yards away was completely surrounded by more water.*

10th tee

*The tee for the 327-yard, par-4, 11th
hole was at the right of the current 16th
green Lower (then the 10th green), fac-
ing the mountain. The green was just
below the current 3rd Upper tee. This
was said to be a rather beautiful hole
and not too difficult though a little hill
climbing was involved.*

11th green

12th tee

12th green

The green for the 210-yard, par-3 13th was the green for the current 15th Lower—at that time the end of the club's property. Of the three par-3 holes, this was said to be the most difficult.

13th tee

13th green

The tee for the 357-yard, par-4, 14th can still be seen between the current 15th green and 14th tee Lower. The hole was about the reverse of the current 15th Lower. Out of bounds and dense woods lined the right side all the way to the green. This hole, 357 yards, par-4, was sometimes called "Sandy's Garden," because of a garden just short of the green. The terrain was very rough, and an abundance of heather took its toll.

14th tee

14th green

The current back tee of the 15th Lower was the tee for the old 15th—in the opposite direction! Two sets of traps confronted short hitters on their first and third shots. Dense woods and out of bounds on the right, the same as on the old 14th, continued to be a major attraction all the way to the green. Contours of the green of this par-5, 450-yard hole are still intact at the far right of the current 17th green Lower.

The 520-yard, par-5, 16th was much the same as the current 18th Lower, but without the elbow—the green being that of the current 3rd Lower.

15th tee

16th tee

16th green

The tee for the 322-yard, par-4, 17th was between the pro shop and the current 3rd green. The second shot was across Baltusrol Way, which was then a public road through the club's property and up the mountain. The green, framed by dense woods, was at the beginning of the 18th fairway Upper on the left and is clearly visible today.

17th tee

The 18th hole, a 275-yard, par-4, was a shortened version of the current 18th Upper and quite easy with one exception. According to Bob Finney, "There was a great big trap in front of the green that had railroad ties on the side toward the green. If you got in the trap, to shoot toward the pin, you had to hit directly at the railroad ties. Why someone playing out of the trap was never killed, I will never know."

18th tee

18th green

The New Courses

In 1916 founder Louis Keller said the club needed two 18-hole courses and offered to lease or sell it additional land. As indicated before, the membership decided in late 1917 to purchase all the land and A. W. Tillinghast, a noted golf architect, was engaged in 1918 to design and construct the two new courses.

Tillinghast was a native of Philadelphia, but he resided in Harrington Park, New Jersey. Ultimately, he was identified with nearly 100 courses, including Winged Foot, the original Philadelphia Country Club, and San Francisco Golf Club, Ridgewood, and Shackamaxon.

According to Frank Hannigan of USGA and golf writer Red Hoffman, Tillinghast, in addition to his genius for building golf courses of quality and character, had other notable attributes.

He was a talented golf player who competed in four U.S. Amateur Championships between 1903 and 1912 and who finished twenty-fifth in the 1910 U.S. Open Championship.

He was a nationally recognized golf writer credited with adding the word *birdie* to the golf lexicon and a contributor to and editor of *Golf Illustrated*, a leading golf magazine, which was published from 1913 to 1934. He also wrote golf fiction and golf humor.

He was a pioneer photographer of golf portraits and golf scenes of professional quality although he considered the activity only a hobby.

He was one of the founding fathers of the Professional Golfers Association of America—the PGA.

He was an early promoter of a professional golf tournament—the 1912 Shawnee Open, which was held at the Shawnee Country Club at Shawnee-on-Delaware, Pennsylvania, the first course he designed. It opened in 1910.

He was one of the first to advocate development of public golf facilities. Indeed, one of his most nota-ble works is a four-course complex at Bethpage State Park in Farmingdale, New York.

He was a champion of the fledgling USGA Green Section and the research and agronomic work it did on turf development.

He was a red-haired dandy with a bristling wax-tipped mustache who during the roaring Twenties was a familiar figure on New York's "Great White Way" as an "angel" of a number of unsuccessful Broadway plays.

Tillinghast completed the Baltusrol Upper and Lower Courses at a cost of $132,000 (exclusive of the cost of the land) in June 1922, four years after he had been commissioned to build them.

The intervention of World War I, of course, was the main reason that the new courses took four years to build. They spread over most of the 470 acres that now comprise the club's property and occupy nearly three times as much ground as the 6,189-yard Old Course, which had a par of 37-37—74.

Despite the prolonged period needed to complete the golf courses, play was never interrupted, according to Robert Finney, a member of the club for more than 60 years and a witness to its transformation. He credited Tillinghast with great ingenuity: The architect had to juggle his work schedule since all of the land occupied by the Old Course was eventually utilized for the two new courses. Finney added that traces of the Old Course can still be found as the routing of both the first and second holes of the Lower; and the last 270 yards of the eighteenth Upper are unchanged. The third of the Upper is approximately the same hole it was before. Besides the first and second holes of the Lower, the current fifth, fifteenth, and sixteenth greens are similar to what they were before Tillinghast worked on them. His modification of the hazard surrounding the sixteenth green was interesting. He replaced a moat with a girdle of sand. Also, he eliminated half a dozen holes that alternately played

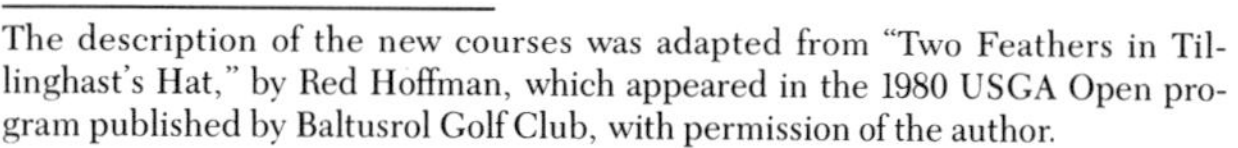

The description of the new courses was adapted from "Two Feathers in Tillinghast's Hat," by Red Hoffman, which appeared in the 1980 USGA Open program published by Baltusrol Golf Club, with permission of the author.

up and down the side of the nearby Baltusrol Mountain at right angles to the general east–west deployment of the current courses.

Of course, occasional changes have since been made in various holes of the courses that were built by Tillinghast in 1922. Here are two examples:

Tillinghast placed the tee of the ninth hole Upper at the back right of the current eighth green Upper instead of to the left, where it is today. And the ninth fairway ran close to the right property line.

Also, on the fourteenth Upper, men played from the current women's tee, and the green was to the right of the current green. What is now a dry ditch was then a small (and wet) dam in front of the green.

The Upper

Except for the expansion and extension of the teeing areas, the relocation of the fourteenth green, and the relocation of the ninth tee and fairway, the Upper is relatively unchanged since it was completed in 1922. Situated as it is on the higher, more rugged terrain of the Baltusrol land, play on it is definitely influenced by the natural elements of its setting, specifically, the 200-foot-high Baltusrol Mountain, which flanks the entire northern boundary of the course.

Obviously, Tillinghast utilized this influence to create a course of implicit strategic design, which forces players to determine where and how they have to play every shot—especially a putt—to compensate for the unrelenting influence of the mountain.

The architect imposed a minimum of bunkering on the fairways of the Upper. And he intelligently incorporated the slope of the ground as a dictate to where tee shots should be played. This is conspicuously evident on the first six holes, which run along the base of the mountain before the course moves onto terrain that does not have such a strong pitch and roll. The greens in this stretch are a golfer's nightmare because the slope of the surfaces seems to go one way and a putt invariably breaks the other way. The 190-yard third hole is an example. Its green is an elongated punchbowl that falls away from a high-faced bunker at the right entrance. Only a perfectly executed shot with plenty of backspin can be expected to hold the green within a reasonable distance from the flagstick. Any indifferent effort is guaranteed to roll to the extreme left behind the green or off the putting surface entirely.

The greens of the Upper are ample although they are generally less expansive than those of the Lower. They are more strongly contoured, with much more restricted entrances. Also, most are of a plateau or pedestal structure and situated so as to have a "right" or "wrong" side from which to putt. When the course is set up for championship play, the right side is certain to be a limited area behind a bunker of consequence, and the wrong side will confront a golfer with a putt of considerable distance across conspicuous surface undulations, plus the unseen, omnipresent influence of the mountain.

From its current championship tees (1984), the Upper is 6,714 yards long with a par of 36-36—72. Its rating by the Metropolitan (N.Y.) Golf Association is 73.2. This is a true reflection of the honest character of the course, which was imparted to it by its designer.

The Lower

The more renowned Lower Course is longer and stronger. As played during an Open Championship, it stretches to a maximum of 7,076 yards, with a par of 34-36—70. It has a USGA rating of 75.8, one of the highest in all golf. Jack Nicklaus described the course as "marvelously fair yet an exceptional test of golf."

Contrasted with the Upper, the Lower has a wide-open look, its greens being larger and much more open to approach.

Its superb greens are not as rife with undulations as those of the Upper, but they are replete with subtleties and nuances almost as mystifying as the mountain's influence on the Upper. Ironically, the slightest, subtlest influence of the mountain is found on the sixteenth and seventeenth greens of the Lower, not the Upper.

The Lower, in preparation for the 1954 Open Championship underwent a considerable facelifting at the hands of Robert Trent Jones, but it still is unmistakably a Tillinghast creation. Jones, a personal friend of the original designer, went to considerable effort to maintain the "Tillinghast touch," explaining:

"In remodeling any golf course of quality, the object is to complement the work of the original. Any major changes should be in keeping with its style. What I did at Baltusrol, I believe, was faithful to the Tillinghast concept."

Many of the changes effected by Jones involved the lengthening of the course by approximately 500 yards to make it commensurate with the high-powered game that had evolved since Baltusrol last hosted the Open. In stretching the course to more than 7,000 yards, he added a number of significant fairway bunkers to emphasize the need for tee-shot accuracy. He also consolidated or enlarged sprawls of strategic sand near or abutting the greens and he eliminated bunkers that were no longer of consequence.

Jones's most conspicuous changes were made to the beautiful fourth hole and the dogleg seventh. In both cases he enlarged the greens to receive longer shots. The fourth was increased by nearly 50 yards to its current 194 yards, and the seventh was reduced from a short par-5 to a maximum par-4 of 470 yards.

In designing the fourth hole, Tillinghast used a pond that requires an absolute over-the-water carry to the green. Jones enhanced and strengthened the hole with a longer carry, a terraced green, and a narrow extension of the putting surface backed by bunkers. The ultimate result is a hole that was selected in 1965 by *Sports Illustrated* as one of the "18 Best Holes in America," and in 1976 by *The World Atlas of Golf* as one of 18 Classic Holes selected from courses on every continent where the game is played.

The "famous fourth" was also the scene of an ultimate squelch by Trent Jones. After remodeling the hole, he was criticized for making it too difficult.

"Let's go play the hole and see if there is anything that needs to be done," Jones suggested as he led the critic along with C. P. Burgess, general chairman of the 1954 Open Championship, and pro Johnny Farrell to the fourth tee, where each struck a shot.

After each of the first three had put his ball on the green, Jones played his—and sank it for a hole-in-one.

"Gentlemen, I think the hole is eminently fair," the architect is reported to have said. (Actually the fourth has been the scene of several notable catastrophies: Arnold Palmer dumped his ball into the water during the 1967 Open. Augie Kammer, the Baltusrol great, took a 9 on the hole in the 1920s. And another member, Henry Topping, vented his particu-

Early fairway equipment

lar frustrations by throwing his entire bag of clubs into the pond.)

Despite the modernization, the Lower retains the flow and flavor of Tillinghast's inspired input, not the least of which was his audacious windup with back-to-back par-5 holes. Although each plays longer than originally—the 630-yard seventeenth with a Sahara of sand for a cross bunker and the 542-yard eighteenth, which funnels through trees and across water—they still warrant the high praise accorded them by the immortal Bobby Jones, who regarded them as "one of the great finishes in golf."

Maintenance

Turf. Maintaining the Baltusrol golf courses in good playing condition throughout the playing season and particularly in August has always been complicated by the fact that Baltusrol Mountain provides a heavy underground water seepage as well as prolonged heavy surface runoffs. Because of these topographic fea-

tures, the location of Baltusrol is "unfortunate," said nationally known agronomist O. J. Noer, who analyzed the club's turf problems in 1938. "Flooding in winter or early spring before growth starts does little damage," Noer stated, "but midsummer flooding for any period is usually fatal. The grass rots out or is killed by other afflictions."

Baltusrol green committees and green superintendents had long struggled with the club's aggravated turf problems. But in the years following the 1967 U.S. Open, extensive drainage facilities were installed throughout the course; indeed, drains were provided for every section of the course where water accumulated. Also, an underground watering system was installed throughout in 1969–1970. And various types of grasses were tried.

At one stage all the fairways were completely plowed under and reseeded in a dramatic shift from predominantly Kentucky bluegrass and Fescue to bentgrasses.

Much of this was done under green chairmen Al-

Later fairway equipment

len Grogan and Robert Finney and green superintendent Joseph R. Flaherty who succeeded Ed Casey in October 1967. Casey had been green superintendent from 1944, when he took over from Major Jones.

According to Joe Flaherty, the grasses now on the fairways of the golf courses are Astoria and seaside bents.

Those on the tees include seaside and Penncross bents.

The greens have several strains of bentgrasses, including Arlington, Washington, and metropolitan. Some greens, particularly on the Upper Course, have isolated patches of velvet bent, which can be identified by its dark green color and very fine leaves.

In addition to the bentgrasses, all greens have a proportion of *Poa annua*, as do all areas of the courses. *Poa*, of course, has made a strong comeback since tricalcium arsenate was banned in 1974.

Baltusrol is actively supporting the USGA's turf improvement program. It assesses members $2 each annually to help finance the study currently under way at the University of Texas to develop grasses that can thrive without water. A hundred other clubs have been following suit.

Trees. The tree population at Baltusrol includes the typical northeastern hardwoods and conifers — including various species of oak and maple; ash; white, red, and black Austrian pine; hemlock; beech; gum (both sweet and sour); locust; many dogwoods, both white and pink; and flowering crabapple and cherries.

The most notable trees on the courses are American elms, the large one on the left of No. 5 Upper fairway, several on the right of No. 6 Upper tee, the large one with the beautiful canopy between No. 5 and No. 17 Lower greens, and the two adjacent to the service road between No. 15 and No. 16 Lower fairways.

Also, there is the copper beech across the brook in front of No. 10 Lower tee, and the one between No. 5 and No. 17 Lower greens.

The evergreens that club manager Major R. Avery Jones obtained from the bankrupt nursery and planted during the middle 1930s form the grove behind No. 3 Lower green (to protect players on that green from errant shots from No. 1 tee), the groves between No. 5 and No. 18 Lower fairways and No. 7 and No. 11 fairways, also to the left of No. 11 blocking the shortcut to the green, to the right of No. 15 pro-

Joseph Flaherty with Executive Vice-President Robert Potter

tecting No. 17 tee, and between No. 18 Lower and No. 18 Upper.

On the Upper, Major Jones planted trees between the 1st and 18th fairways, to the left of No. 7 green, to the right of No. 10, between No. 4 and No. 16, to the left of No. 5 fairway, and between No. 6 green and No. 12 tee.

An arbor committee was formed in 1975 to replace lost trees, to back up trees that are essential to a particular hole, and to add color to the stands of evergreen, many of which have been damaged by disease. The program is financed by voluntary contributions by members, which have totaled over $65,000.

Around 3,300 trees of many varieties were planted by the committee after careful study and advice from consultants. According to Robert Potter, vice president of the club in 1985, the program has finally caught up with the persistent attrition of the club's lovely foliage. The trees planted since 1975 include:

Japanese cherry tree	Hawthornes
Ginkgos	Witch hazel (Chinese)
Mountain ash	White pines
Hemlocks	Willows
Lindens	Calley pear
Dogwood	Flowering crabs
Olive	Elms
Locusts	Oaks
Katsurba	Blue Atlas cedar
Maples	Sweet gum
Dawn redwoods	

Ed Casey former greenkeeper with members

CONTEMPORARY SCENES: A COLOR PORTFOLIO

On the 18th lower

The 18th lower in the shadows

Approaching 5th lower

New pro shop—former President Robert Boutillier, pro Bob Ross, and Chuck Simpson

13th lower green

2nd lower green

On the 11th lower

Terrace Lounge—1984

Cocktail Lounge—1984

Bob Ross, head pro, Jean Verdi, chef

Dining Room — 1984

Mens Locker Room Solarium

15th lower

16th lower

18th upper

9th lower

1st lower

3rd lower

15th upper, looking towards tee and 4th green

Famous 4th lower—two views

18th lower fairway

2nd lower

9th upper

11th upper

8th upper

Approaching 4th lower

Approach—5th lower

Tee—7th lower

10th lower

8th lower

15th lower

On the upper at dusk

From the 4th lower green

NATIONAL
CHAMPIONSHIPS
PLAYED AT BALTUSROL

From its very inception, Baltusrol Golf Club has been in the mainstream of American golf. No club has taken its responsibilities to the sport more seriously. Its courses have always been open to major tournaments despite the inevitable inconveniences to members that go with hosting such events—not to mention the immense amount of work that goes into mounting even a single U.S. Open. Notwithstanding, Baltusrol has accepted many of these tournaments.

The top golfers of every generation have played at the club at one time or another during its 90-year history, including Willie Anderson, Harry Vardon, Edward Ray, Alex Smith, Francis Ouimet, Walter Hagen, Jerome Travers, Charles Evans, Gene Sarazen, Bobby Jones, Johnny Farrell, Byron Nelson, Ben Hogan, Craig Wood, Cary Middlecoff, Gary Player, Billy Casper, Arnold Palmer, Gene Littler, Jack Nicklaus, Lee Trevino, Johnny Miller, Hale Irwin, and Tom Watson—to name just a few who won the U.S. Open—and the great Sammy Snead, who never did win it. And there were the great amateurs—Walter J. Travis, Harold Hilton, H. Chandler Egan, Max Marston, and Johnny Goodman.

Top women golfers have played at Baltusrol too, among them Frances C. Griscom, Beatrix Hoyt, Ruth Underhill, Margaret Curtis, Betsy Rawls, and Mickey Wright, all of whom have won the U.S. Women's Open or U.S. Women's Amateur Championships.

These illustrious names and many others, both men and women, appear in the program descriptions of the 12 national championships that have been played at Baltusrol. To read these descriptions is to read much of the history of American golf. For just as every great opera star ultimately finds his or her way to the Metropolitan Opera, every great golfer ultimately performs at Baltusrol and but a few other mainstream clubs that are similarly dedicated to the sport.

Descriptions of the national tournaments were adapted with the permission of the authors, from the following articles, which appeared in the 1967 and 1980 USGA Open programs published by Baltusrol Golf Club:

1901, 1903, 1904, 1911, 1915 (part), and 1946 tournaments — From "They've All Played Here," by Charles Price, 1980 program.

1915 (part), 1926, 1936, and 1961 tournaments — From "Champions Are Unpredictable at Baltusrol," by Des Sullivan, 1967 program.

1954 tournament — From "Out of the Rough in 1954," by Red Hoffman, 1967 program.

1967 tournament — From "Jack Will Be Back," by Kaye Kessler, and "Last Time Around," by Frank Hannigan, 1980 program.

The description of the 1980 tournament was written by Joe Miegoc of USGA.

1901 USGA Women's Amateur Championship

The seventh U.S. Women's Amateur Championship was played at Baltusrol on October 8–12, 1901, before a gallery of nearly 1,000. It was won by Genevieve Hecker (who later became Mrs. Charles Stout). Dressed in the fashion of the day, Miss Hecker wore a sailor hat, a starched collar, a blouse with sleeves the size of watermelons, and a wasp-waisted skirt lined with leather at the hem to protect it from fraying as it ruffled the turf.

There were 84 entries with qualifying scores ranging from 97 to 104, including 7 women from Boston who "created consternation by their steadiness of play." The preceding year's champion, Philadelphia's celebrated Frances Griscom, had failed to qualify— she took 4 putts on each of 7 greens.

Miss Hecker, then a tigress in the New York Metropolitan golf district, defeated Lucy Herron of Cincinnati, 5 and 3, in the final. Miss Herron, whose short game deserted her, had defeated Margaret Curtis in the semifinals. But Miss Curtis was the medalist —and she would go on to win the U.S. Women's Amateur Championship three times, in 1907, 1911, and 1912. The 1911 tournament was held at Baltusrol.

Miss Hecker, who won the Women's Metropolitan Golf Association Championship in 1900, 1901, and 1906, successfully defended her 1901 U.S. Amateur title the following year at The Country Club in Brookline, Massachusetts. She has gone down in golf lore as the first woman to write a book on golf, entitled, simply enough, *Golf for Women*, which she described in a subtitle as being aimed at "the feminine, inquiring mind and from a woman's point of view." More about Genevieve Hecker later.

Genevieve Hecker

1903 USGA Open Championship

The second USGA Championship to be staged at Baltusrol was the 1903 Open on June 26–27. The legendary Willie Anderson won it in a playoff with David Brown, a former British Open champion.

Anderson had been the pro at Baltusrol for about a year, having been hired by Louis Keller in 1898 mainly because he had been born in Scotland and because no self-respecting club then considered itself a real golf club unless it had a Scottish pro.

What Keller may or may not have known was that Anderson had never hit a serious golf shot in Scotland. He had mysteriously picked up the game in America—somehow. He ultimately won four Western Opens and was the first man to win four National Opens. This record has since been matched only by Bobby Jones, Ben Hogan, and Jack Nicklaus.

The field in the 1903 Open played 36 holes each day. Anderson started with 73-76—149, the 73 then being a record for the Open. At that point he had a 5-shot lead, and Brown was 7 shots behind. But the second day Anderson shot 76-82—158 and was tied with David Brown at 307. The 82 was attributable to an 8 on the par-3 third hole of the Old Course, Anderson having hooked his ball into the rocks with his spoon.

Johnny Farrell, former Baltusrol pro, likes to retell the experience of one of the club's oldest members who had watched Anderson's tribulations on the third hole. Anderson said to the member: "Would you like to have this spoon?" The member took the club and cherished it. "Imagine winning the Open with an 8 on a par-3 hole," chuckled Farrell.

The playoff was delayed by a downpour at 10 a.m., with some greens almost submerged. Brown was opposed to playing at all, but Anderson was ready, and a little before noon, Herbert Windler, USGA president, ordered them to go ahead. Anderson led by 2 at the ninth; they were even at the fourteenth; and Anderson picked up a stroke on each of the fifteenth and sixteenth holes and held the lead to win.

Willie Anderson and Alex Smith

1904 USGA Amateur Championship

In September 1904, Baltusrol held its third national championship, this time the U.S. Amateur. It was won by H. Chandler Egan, a handsome blond lad of 20 from Chicago, who was still a student at Harvard.

Egan defeated Fred Herreshoff, 8 and 6. Fred was a nephew of Nat Herreshoff, a famous boat designer of the time. Tillinghast, who built today's Lower and Upper Courses at Baltusrol some 15 years later, participated but was beaten in the first round by Egan. Jerome D. Travers of Upper Montclair Country Club, who later won the National Amateur four times, was also eliminated in the first round. And Walter Travis, an American and a previous three-time winner of the U.S. Amateur—1900, 1901, and 1903—and runner-up in the 1902 U.S. Open, was eliminated in the second round.

According to H.B. Martin's *Fifty Years of American Golf*, the dignified, black-bearded, and taciturn Travis, who was 42 years old, had just stunned British golfdom by winning the 1904 British Amateur Championship at Sandwich. His successful invasion of the United Kingdom was one of the most memorable events in American golf—"doing more than anything else to promote the game on this side of the Atlantic." The British were reportedly upset and wholly unprepared for such a catastrophe. In presenting the cup to Travis, the obviously piqued Lord Northbourne is quoted as saying: "Never since the days of Caesar has the British nation been subjected to such humiliation."

But back to Chandler Egan.

Egan, whose real first name was Henry, was one of the most impressive and enigmatic players in early American golf. He won not only the Amateur title at Baltusrol in 1904 but the qualifying medal as well—and he won the championship again the next year at the Chicago Golf Club, where he was also runner-up in 1909. Then in 1911, for reasons known only to himself, he moved to Oregon to raise fruit on a farm that was 300 miles from the nearest golf course.

It was in 1929, twenty years after he had last played in the Amateur Championship, that Egan again decided to enter. He was then 45, which would be 60 by today's geriatrics, and he had limited his scant tournament appearances to the Pacific Northwest. Notwithstanding, he beat three nationally ranking, vastly younger players before being put out in the semifinals. Ironically, for a man who could take his golf or leave it, he caught pneumonia on a golf course seven years later and died within a week.

H. Chandler Egan

*Walter Travis,
Jerome Travers—1912*

1911 USGA Women's Amateur Championship

On October 9–14, 1911, the women came back to Baltusrol for their National Championship, and this time it was won not surprisingly by Margaret Curtis, who defeated Lillian Hyde, 5 and 3. Mrs. R. H. Barlow of Philadelphia was the medalist with 87.

Margaret Curtis had won the Amateur in 1907 at Midlothian and would win again in 1912 at the Essex County Club, her home course near Boston. Her older sister, Harriot, had won the title first in 1906, and it had been Harriot whom Margaret had defeated in the final in 1907, this having been the one and only occasion on which two sisters have met in the final of a national golf championship. All told, Miss Margaret, as she came to be known in golf circles, played in 25 USGA championships—her first was in 1897 and her last was in 1949, when she was 65.

On May 25, 1905, both Curtises, Margaret and Harriot, played an informal match against leading British women golfers at Royal Cromer in England. So pleasurable did they find the contest that they decided to somehow institute a regular tournament involving women golfers from many lands. However, the first contest did not take place until 1932. But since then, except for the years during World War II, the tournament has been held every two years and is known, appropriately enough, as the Curtis Cup Match.

Incidentally, reporting on the 1911 Women's Amateur at Baltusrol, *The New York Times* observed, "The course is generally regarded as one of the sportiest in the East."

Margaret Curtis

1915 USGA Open Championship

Jerome D. ("Jerry") Travers of the Upper Mont-clair Country Club dominated amateur golf in 1915 as Bobby Jones would later dominate the professional field as well. Travers won the National Amateur Championships in 1907, 1908, 1912, and 1913. He also won the New Jersey Amateur three times and the Metropolitan Amateur five times. Actually, Travers was one of the three kingpins of amateur golf, the others being Charles E. ("Chick") Evans of Chicago, who would win both the National Amateur and Open titles in 1916, and Boston's Francis Ouimet, who at age 20 had become a national golfing hero by beating the fearsome Englishmen Harry Vardon and Ted Ray in the 1913 Open at Brookline. But Travers had never made an impression in the U.S. Open.

With 18 holes to play in the final round of the 1915 Championship over the Old Course at Baltusrol, Travers, to his apparent amazement, found himself with a 1-stroke lead. By the final turn, he also discovered he had to negotiate the last nine holes in 1 under par in order to finish ahead of Tom McNamara, a leading Boston pro who had been runner-up in two previous Opens, 1909 and 1915.

Distrustful of his driver, Travers turned to his heavy driving iron on the tenth hole and promptly sliced out of bounds. His next effort was an equally wild hook, but a magnificent pitch from the rough to within a few feet of the cup gave him a sweating par. (Out of bounds in 1915 meant loss of distance only.)

Frustrated, Travers again tried his driver, only to top the ball into the rough. He salvaged his par with two long irons to the green, went back to the driving iron, and finally bagged his elusive birdie on the long fifteenth by successfully gambling on a herculean carry over a cross-bunker.

Then, regarding par as the equivalent of the Holy Grail, Travers played the last three holes with solid precision for a score of 297, a highly respectable total in those days. He won by a stroke over Tom McNamara, by 2 strokes over Bob MacDonald, another professional, and by 4 strokes over Frenchman Louis Tellier and Long Jim Barnes, a former Cornishman, who would later become an assistant to Johnny Farrell at Baltusrol.

Jerome D. Travers

1926 USGA Amateur Championship

I n 1926 one golfer towered above all others. Robert Tyre Jones, Jr., of Atlanta, Georgia, occupied a place in the sport later approached only by Hogan, Palmer, and Nicklaus. After scoring his first major victory in the 1923 U.S. Open, Bob Jones captured the National Amateur in 1924 and 1925. Not long before he came to Baltusrol (in September 1926) to defend that title, he had won the British Open for the first of three times and the U.S. Open for the second of four.

Playing on Baltusrol's Lower Course in the 1926 U.S. Amateur, Jones had a comparatively comfortable route to the final—but none of his victories, including triumphs over his archrivals, Chick Evans (3 and 2) and Francis Ouimet (5 and 4), were walkovers. Nevertheless, there appeared little doubt as to the outcome of the championship—George Von Elm, the blond Uhlan from California, would serve merely as the last steppingstone on Jones' exalted progression to an unprecedented third consecutive U.S. Amateur title. Von Elm had succumbed easily to Jones in two previous meetings, losing 9 and 8 in the 1924 final and 7 and 6 in the 1925 semifinals.

But Von Elm surely was a warrior who had fate nailed to his shield that week. He barely survived the 36-hole qualifying test, taking 16 strokes more than medalist Jones, and went an extra hole in his first match after narrowly escaping elimination on the home green.

But in the final, the Westerner gave the Southerner no openings, and Jones was relegated to playing defensive golf.

Von Elm was 1 up at 9, at 18, and at 27. He won the twenty-eighth to go 2 up. On the thirty-first hole, the thirteenth Lower, Jones pushed his drive into the ditch, played out of the water into a trap at the green, and wound up with a 5. With Von Elm 3 up, Jones won the fourteenth when Von Elm 3-putted. They halved the seventeenth, and Von Elm won 2 and 1.

A gallery of 12,000 saw the quarterfinals, after which the ticket prices were increased from $1.10 to $1.20—*The New York Times* observing that it was becoming as expensive to watch golf as to play it.

E. M. Wild, nine-time club champion at Baltusrol, qualified for the 1926 Amateur but was eliminated in the first round.

George Von Elm

Von Elm and Bobby Jones—1926 USGA Amateur Championship

1936 USGA Open Championship

I n 1936 the Upper Course was the scene of a most unpredictable Open victory in which two new records were set within 20 minutes.

Easily to be expected would have been a triumph by "Lighthorse" Harry Cooper, regularly among the leading money winners of that post Depression era. After weathering threats from Ray Mangrum, Paul Runyan, and Vic Ghezzi, Cooper turned in a score of 284 for seventy-two holes, 2 strokes below the Open record set by Chick Evans in 1916 and equaled by Gene Sarazen in 1923.

It was hard for Cooper to fend off the congratulations showered on him as he walked to the locker room. But he was not a happy man. He had spoiled what could have been a great round by tailing off in the stretch, taking bogeys on the fourteenth, fifteenth and eighteenth holes. Yet his 1-over-par 73 was a solid performance at the end of a long tournament.

But suddenly the crowd of well-wishers began to evaporate when almost-impossible-to-believe rumors filtered into the club house about Tony Manero, a dapper little ex-caddie, almost unknown beyond the boundaries of the New York metropolitan area. Manero, who was playing with Gene Sarazen, had started the final round four behind Cooper but was 5 under par after 13.

Reporters, who had begun to type their stories hailing Cooper as the successor to Sam Parks, Jr., the 1935 darkhorse winner, reluctantly left their machines and plodded wearily to join the throngs now gathering around Manero at the fifteenth green. They saw a tense, nervous Tony trap his short iron for a bogey 4. But they watched expectantly, knowing a new lead most certainly would have to be written within minutes, when Manero dropped a 12-footer for a 3 on the 439-yard sixteenth to go 5 under regulation figures again.

Tony needed two pars to win by 2, and he carded them both in almost routine fashion, scoring 67 for the round and 282 for the tournament—another new Open record. Tony never again was a factor in the Open, although he reportedly won thirteen other tournaments against pretty much the same opposition.

The press coverage of the 1936 Open, reportedly an estimated half-million words, was the maximum up to that time. But the members of the press were not happy. Club manager Major Jones had converted sleeping quarters over the garage into a press room, and they turned out to be hot enough to earn the scurrilous title of "The Black Hole of Calcutta."

Tony Manero

1946 USGA Amateur Championship

The 1946 U.S. Amateur was the first since 1941, the tournament having been suspended for four years during World War II. Baltusrol's Lower Course was the scene, and the championship was won by Stanley E. ("Ted") Bishop, New England and Massachusetts state amateur champion. He defeated Smiley L. Quick, a U.S. Public Links champion from California. Bishop had been a professional for a short period before the war but had regained his amateur status.

The field was replete with "name" amateurs, some of whom would go on to become noted professionals. Included were Charles Evans, Jr., who had played in every U.S. Amateur since 1907—a span of 39 years. He won the Amateurs in 1916 and 1920 and was runner-up in three others. And he won the U.S. Open in 1916.

Also entered was Max Marston, who had been Baltusrol club champion in 1914, 1915 and 1916 and who had won the U.S. Amateur in 1923 and was runner-up in 1936.

A.F. Kammer, Jr., who was Baltusrol club champion in 1930, 1936, 1937, and 1938, was also there. Kammer, whose father had also been a Baltusrol great, beat Cary Middlecoff in the quarterfinals of the 1946 tournament but was eliminated in the semifinals by the ultimate runner-up, Smiley Quick. Middlecoff went on to win the U.S. Open in 1949 and 1956.

Others who played in the 1946 Amateur were John W. Fisher, 1936 U.S. Amateur champion; Johnny Goodman, 1937 Amateur champion and 1933 Open winner—the fifth and last amateur to win the Open; Willie Turnesa, 1938 and 1948 Amateur champion; Bud Ward, 1939 and 1941 Amateur winner; and Dick Chapman, who won in 1940.

Still others were Doug Ford and Frank Stranahan along with Robert H. ("Skee") Riegel, who later won the 1947 Amateur, and Dick Mayer, who later won the 1957 Open.

In the 1946 final at Baltusrol, Smiley Quick was 2 up on Bishop at 18 and even at 27, and they halved 9 holes in a row. On 17 Bishop had holed a 25-foot putt and Quick a 10-footer, but both missed with less than perfect putts on the thirty-sixth. On the thirty-seventh hole, Bishop chipped up and sank a 4-footer and Quick missed from 2½ feet.

At one stage, according to Johnny Farrell, Quick demanded that newsreel cameras be stopped behind the fifth green, saying "He [Bishop] can't hear them; he's deaf. But I can."

Stanley E. Bishop

1954 USGA Open Championship

The field in the 1954 Open left little to be desired: defending champion Ben Hogan, along with Sam Snead, Bobby Locke, Cary Middlecoff, Lloyd Mangrum, Julius Boros, Doug Ford, Jackie Burke, Jimmy Demaret—the greatest players in the game. But none of them won it. Instead it was a lame-armed pro from Clayton, Missouri, who only shortly before had settled down to a club job after ten frustrating years on the PGA Tour during which he had achieved but one tournament victory. He was Ed Furgol, who for three days was the greatest player in the world.

His victory over the field of 161 brilliant golfers and over his physical disability has to be one of the greatest in the annals of the Open, but Ed would be the first to agree that there was some element of luck to his conquest.

To anyone in the record crowd of 36,909 attending the Open, and to the millions who saw it on national television for the first time, it is easy to recall

Ed Furgol

how Furgol played two golf courses on the final hole —
from the woods of the Lower to the eighteenth fairway
of the Upper and back again to the apron of the eighteenth green Lower to get his par and the title. But
not many remember how Lady Luck smiled on him a
few minutes before at the seventeenth hole, the seventy-first of the tournament.

On that 620-yard stretch, one of the longest ever
to confront a golfer in any Open, Furgol was in dire
straits when he sliced his tee shot into deep rough behind the fifteenth tee. Confronting him was an even
more formidable problem, the "Sahara" bunker,
sprawled completely across his line of play to the
green. He was 140 yards from the front of the hazard,
but he needed a carry of 170 yards to fly it.

Ed dug the ball out with a 4-iron, but the heavy
grass cushioned the blow and the ball fell short—in
the bunker.

It went into that wide expanse of sand, but it did
not stay there. Instead, as if guided by radar, it rolled
on a relatively straight line along a path of hard-
packed sand that cut through the entire bunker. The
ball rolled and rolled and never stopped until it
reached the fairway beyond. For Furgol a 5 was easy
from there. He played a medium iron to the green and
2-putted for his par.

Dick Mayer

Mountain Tricks Littler

Gene Littler chased the champion during the long final day of play only to miss an 8-foot putt for a birdie 4 on the seventy-second green that would have enabled him to catch Ed. That putt alone did not prevent the 1953 National Amateur titleholder from overtaking his quarry. Actually it was a missed 3-footer on the sixteenth green, where he was victimized by the most insidious roll on the entire golf course.

Littler was even with Furgol after sixty-nine holes, and he appeared likely to maintain his position when he recovered to within a yard of the cup after flying his tee shot into a greenside bunker at the sixteenth, a 201-yard par-3. Gene had what appeared to be a straight-in putt for his par. And that is the way he played it, but the ball did not drop. It edged the hole and stayed out. Littler took a 4 and fell behind by 1 stroke.

Littler's eyes had not played him tricks, but Baltusrol Mountain 150 yards to the north had. Nowhere on any of the thirty-six greens on Baltusrol's two courses is the effect of that rise of ground as subtle. But the break is there, and it cost Gene his most vital par—and it deferred until 1961 his winning of golf's greatest championship.

Two 7s Finish Mayer

Dick Mayer stayed within sight of Furgol even longer—almost to the clubhouse—only to lose him on the seventy-second. Handsome Dick, who won the Open three years later at Inverness, is credited with opening the door for Furgol by taking a 7 on the final hole when a par-5 would have put him in a playoff.

Actually, Dick could have sliced deep into the evergreens on the 545-yard finisher, played a provisional ball—as he did—and taken his 7 and still won had he not posted a 7 earlier. That first 7 eliminated the cushion he needed to survive the second.

Mayer messed up the relatively short 379-yard, par-4, second hole in his third round. Playing safe, Dick tried to fade a 4-wood tee shot to avoid going out of bounds. He hooked it instead. The ball caught a "sentinel" tree, which marked an abrupt right angle direction change along the property line of the course, some 200 yards from the tee. Mayer then compounded his error by 3-putting after reaching the green in 4 with his second ball.

Mayer settled down to play some of the best golf of the tournament—he was 3 under for the remainder of the third round to finish with a 70, and he was even for 17 holes in the afternoon, until he went trailblazing among the fir trees.

Hogan–Snead Mishaps

The first hole of 460 yards, which played to a par-4, was the scene of anguish for both Ben Hogan and Sam Snead.

Ben, the defending champion, was seeking an unprecedented fifth Open. The first ball he struck in the second round put him in a hole, literally. It faded over the crowd lining the fairway from tee to green and came to rest in a small gully hidden behind a bunker some 240 yards from the tee. It was a steep declivity with weeds along its sides and water at its lowest point. You could almost hear the wheels turn as Hogan studiously weighed the problems of trying to recover from the sharp upslope of the bank.

Unexpectedly, Hogan got help with his deliberations from the great West Virginia amateur Bill Campbell, who was in the same hazard and "away." Bill made a vain attempt to extricate the ball but succeeded only in dislodging it, and it rolled down into the water.

Hogan, after observing this, turned to his caddie and said: "Pick it up." Ben took the stroke penalty, played a long iron to the green and 2-putted for a bogey.

Snead, even then trying to gain the one major title that had eluded him, found his trouble on the other side of the fairway, deep in the rough. It came as a result of his first—and possibly his worst—shot of the tournament. The sweet swinger hit a screaming duck hook, quail high, which sent the big gallery scrambling for safety. The ball came to rest 18 inches from a wire fence marking the out of bounds, 180 yards from the tee.

"Lawdy me, what a way to start a tournament," Sam moaned when he saw where the ball was. Then, briefly stomping around to determine that there was no way to strike the ball from the right side, he ordered his caddie: "Give me an 8 arn."

Snead then turned the clubhead upside down, took a couple of practice swings, and proceeded to strike the ball left-handed, sending it some 160 yards to the middle of the fairway from where he pitched on and 2-putted for a 5.

Weitzel and Chapman Aces

The 1954 tournament also was noteworthy because it was the first Open in which a hole-in-1 was scored by two golfers, Johnny Weitzel and 1940 National Amateur champion Dick Chapman.

Weitzel, ill-fated successor to Hogan at Pennsylvania's Hershey Country Club, made his in the first round. He drilled a 2-iron over the bunkers on the 190-yard twelfth, where the ball took two bounces on the green and disappeared into the cup. Johnny, killed in an auto accident in 1959, was the first Open tournament golfer to record such a feat in 18 years. Zell Eaton had made the last one in 1936 at Baltusrol, no less, on the tenth hole of the Upper Course.

Chapman coupled his ace with the lowest round of the tournament, a 67 on the second day. The dark-haired internationalist's 3-iron shot hit the front of the green on the 204-yard ninth and rolled into the cup for a hole-in-1. That shot climaxed what was unquestionably the most spectacular stretch of golf in the tournament. Over a span of four holes, Dick scored a chip-in birdie from 60 feet, exploded from a bunker into the hole for another birdie—both 3—and then scored the ace. He was out in 32, two under, and back in 35, one under, 2 strokes lower than any other posted as only 7 golfers broke par during the entire championship.

Patton Low Amateur

Despite his fantastic 67, Chapman was not low amateur. Billy Joe Patton, the fast-swinging lumberman from Morgantown, North Carolina, won that honor. Patton, who two months before had come as close as any amateur ever to winning the Masters, was the first-round leader at Baltusrol, with a 69. He subsequently finished with a 289 total to tie for sixth, 7 strokes better than Chapman, who finished tied with Weitzel for twenty-first.

Billy Joe, who had won as many friends with his quicksilver wit as he had with his derring-do golf game, did not disappoint anybody on either count at Baltusrol.

Upon being presented his medal as low amateur by then USGA president Ike Grainger of Montclair, he sent a glow through the massed thousands when he declared:

"Everyone up here in Yankee land has been so nice. I almost forgot my grandpappy fought at Gettysburg."

Sam Snead

1961 USGA Women's Open Championship

Mickey Wright

Mickey Wright would have been happy to leave Baltusrol and its big, fast, undulating greens to the birds, not to mention to the other women, after she struggled to an 80 over the Lower Course in the second round of the 1961 U.S. Women's Open Championship. Her 36-hole total was 152, four shots off the pace set by Ruth Jessen and Jo-Anne Prentice.

But that was not the kind of a showing Mickey expected of herself, nor was it what the gallery expected of the attractive blond Texan. This young woman, who had been national junior champion in 1952, was being hailed as the only possible successor to the fabulous Babe Didrikson Zaharias, whose death had left a choking void in all of golf. Miss Wright had won the Open in 1958 and 1959, breaking tournament scoring records in each instance. And she would win it again in 1964. Indeed, between 1955, when she joined the LPGA Tour, and 1965, she won not only 4 U.S. Opens and 4 LPGA championships but 57 other tournaments as well, for a total of 65 victories.

The dramatic surge that Mickey produced in the third round on Saturday morning before a record crowd of 3,757 was not expected. She had no fewer than 6 birdies on her card and scored a 69, 2 ahead of Betsy Rawls.

Her afternoon round was a lesson in shotmaking; although she was hitting 3-irons, 4-irons, and 5-irons to the greens, she made only two errors. She pulled her approach into a bunker on the 3rd, and on the short 16th her long iron hung out on the right just a shade and found a bunker. On 16 of the 18 greens, she had eminently makeable birdie putts. She made none of them, 2-putting every green. Had she been holing out then as she had been in the morning when she needed only 28 putts, she could have been in the neighborhood of 66. Instead, she shot 72, but her 293 won the tournament by 6 strokes over Miss Rawls.

1967 USGA Open Championship

Arnold Palmer came to Baltusrol in June 1967 with $91,000 in earnings, the top in the world. Jack Nicklaus came to Baltusrol after failing to make the cut in the Masters only a few weeks before. In addition to a classic duel between these two, the sixty-seventh Open saw Lee Trevino break into the big time and Ben Hogan bow out of Open Championships. The redoubtable Hogan, a four-time winner, was at 54, the second oldest (to 57-year-old Claude Harmon) in the field. He said, "I figure to win any tournament I enter."

An amusing incident occurred on Monday, the first practice day. A spectator with a bag of clubs over his shoulder approached an entrance gate. Matt Glennon, one of the Baltusrol officials, inquired as to the man's purpose. The man said that he had been given a ticket to a practice day of a golf tournament and that he therefore had come to play a practice round. It said "Practice Day" right on his ticket.

On Wednesday Jack Nicklaus fashioned an incredible 62 in a tuneup round in the company of Palmer that quieted some critics' carping about his allegedly crumbling game but that did not, at the time, shake Palmer.

"I can't imagine a round like that shaking anybody up," said Palmer, who paid Nicklaus $15 to see it in a friendly Nassau round.

Nicklaus also dismissed the tuneup round as nothing to be unduly excited about, though he did admit that he had unearthed a new putting stroke with an old friend, a bull's-eye putter he had given a coat of paint and called "White Fang," that allowed him to have nine 1-putt greens. Then he needled Palmer by reminding him playfully, "That's the lowest score I ever shot in the U.S., but I did shoot another 62 the day before the Australian Open of 1964—which I won."

With that background, Nicklaus went out and had himself a delightful time on the course.

Jack put up an opening 1-over-par 71 on a threat-ening day when late storms shook Hogan into a 3-putt at 18 for a 72, a day that belonged to 23-year-old amateur Marty Fleckman, who extracted a 67 out of the historic course. Palmer and Casper would match 69s, and Nicklaus was not unhappy with his 71, even if USGA watchdogs, who the year before at Olympic had warned him for slow play, dogged his every footstep. He played with Bob Goalby and Mason Rudolph in 4 hours and 35 minutes and announced, "We waited every single shot."

Palmer took command the second round with a 68 and 137 total, but Nicklaus had a 67, to take second place a shot up on Casper and 2 ahead of the faltering (73) Fleckman. It was a memorable day in which Rives McBee would exit after rounds of 76 and 72 with the distinction of having had six different caddies—one a newsman, one who worked on his grips in an ornery way, one who sassed back, one who fainted, and one ruled ineligible by the USGA because he had caddied for another golfer in the morning.

1967 Open

Fleckman reclaimed the lead in the third round, but it was a momentary reprieve. Nicklaus followed a troubled 72 on Saturday with an astounding 65 on Sunday for a 4-shot victory margin over Palmer, the runner-up. It was a record 5-under-par 275 total that wiped out Hogan's U.S. Open record, set at Riviera in 1948.

Oddly, Nicklaus had told hometown crony Bob Hoag, "If I win the Open, I'll break the record."

There were eight birdies in Nicklaus's closing 65, the 22-foot birdie putt on the seventh hole putting an end to Arnie's challenge. They came to the seventh with Nicklaus a stroke ahead. Both men hit good drives. Palmer was away. He played a truly magnificent 2-iron that covered the flag all the way, cleared the bunkers and dug into the green 8 feet from the hole. Nicklaus's shot seemed almost indifferent by comparison; it hit the green 22 feet away. If Palmer holed his birdie putt and Nicklaus made a par-4, as seemed likely, Palmer would have caught up. Instead, Nicklaus stroked a putt that was in from the instant he hit it; Palmer missed, and instead of being even, he fell 2 strokes behind.

But the record-setter came at the 542-yard eighteenth. "Once I got the lead on the fourth hole, all I wanted to do was beat the rain," said Nicklaus, who took a 1-iron off the tee on the final hole. "I had two things on my mind. I had a 4-shot lead, and the only way I could be tied was for Arnold to make 3 and me make 7."

With Eastern golf fans screeching, "C'mon, Fatso, duck hook it into the woods" (which is what Ed Furgol had done in the 1954 Open), Jack drew out the 1-iron and pushed it into the right rough. It snuggled under a TV cable and two free drops later he punched an 8-iron short of the water ditch, followed with a thundering 1-iron "maybe 230 yards uphill" that crawled onto the green and probably was the last clap of thunder that touched off the cloudburst.

But not before he rammed in the uphill 22-foot birdie putt for the record.

Trevino Debut

The 1967 Open Championship was memorable more than for Nicklaus's record 275.

It was the advent of Lee Trevino's Open showings. Lee, an assistant pro playing out of El Paso, Texas, finished fifth and won a prize of $6,000. He was the quintessential unknown player who every so often lends a special meaning to the Open. Trevino was not a tour "rabbit" in 1967. He had never played on the Tour. He was a guy from Texas with a strange swing and an appealing manner. Trevino pocketed his $6,000 and allowed as how he might not go back to El Paso but might try the Tour. One year later Trevino broke 70 in every round to win the Open at Oak Hill in Rochester, New York, and tie Nicklaus's 72-hole record of 275.

Lee's 283 to tie for fifth in 1967 was a shot ahead of Deane Beman, who later became PGA Tour commissioner. Beman, the two-time U.S. Amateur Champion (1960 and 1963), had turned pro in the spring of 1967 at the relatively ripe age of 29. The Open was only his second event as a professional. Beman, whose pinpoint accuracy made him a factor in a number of Opens, did something that does not happen very often; he played the first hole in the four rounds in a total of 12 strokes. The first hole is a par-4 of 465 yards. Beman was not a long driver, so he needed a wood for his second shot. On the first day Beman holed his second shot (about 230 yards) for an eagle 2; in both the second and third rounds he made birdies; and in the fourth round he made a routine par-4. Beman shot 284 to tie for sixth place.

The 1967 Open Championship was also the end of the Hogan era in Opens. The 54-year-old shot 72-72-76-72 for 292 and thirty-fourth money of $940, insisting, "Hell can't be any hotter than this."

There was a poignant incident involving Hogan on Thursday. Play had been interrupted by a thunderstorm while Hogan was playing the eighteenth hole. Hogan had stayed on the course. A spectator, Norb Anderson of St. Paul, Minnesota, ducked under the ropes and shared his umbrella with Hogan. Anderson, who had slipped off from a business meeting to see his idol Hogan for the first and only time, was enchanted. Play resumed when the rain abated. Anderson moved to the side but stayed in the fairway as Hogan played up to the eighteenth. He then returned to Hogan's side, covered him with the umbrella, and up the hill to the eighteenth green they went, Ben Hogan and a public links golfer from St. Paul.

Jack Nicklaus

1980 USGA Open Championship

The eightieth U.S. Open Championship, on Baltusrol Golf Club's Lower Course, found Jack Nicklaus's career at a crossroads.

Nicklaus was returning to the course where he had won his third National Open in 1967, with a 72-hole record of 275. But he was also coming off a poor 1979 season. He had failed to win a tournament that year and had earned just $59,434 on the PGA tour, his first season below the $100,000 mark since his rookie year of 1962. His earnings rank of seventy-first marked the first time he had finished below fourth.

But in a remarkable four-day period, Nicklaus proved he was still a force to be reckoned with. He equaled the 18-hole Open record when he opened with a 63, set 36-hole and 54-hole records, and his winning 272 total broke his own previous 72-hole record by 3 strokes. His fourth Open victory matched the record co-held by Ben Hogan, Robert T. Jones, Jr., and Willie Anderson.

But on the first day at Baltusrol, Nicklaus did not have the limelight to himself.

Nicklaus's opening 63, which could have been a stroke lower had he not missed a 3-foot birdie putt on the eighteenth green, was equaled by Tom Weiskopf, and indeed low scores were the order of the day. New Jersey had had 24 inches of rain since the first of the year, and players were throwing shots right at the flagsticks and hold them. In addition to the two 63s, three others shot 66, three more shot 67—in all 19 players broke par, which was 70.

By the time the championship was over, 49 rounds were played under 70. Tom Watson, who scored a hole-in-1 on the fourth hole during the championship, finished with a 276 total, a score that would have won or at least gained a playoff in all but 2 of the previous 79 Opens. Watson finished fourth, in a tie with Lon Hinkle and Keith Fergus.

On that sunny, windless first day, Nicklaus went out in 32 with three birdies and a bogey on the 377-yard second hole. He came back in 31, with three con-secutive birdies beginning at the eleventh hole, along with birdies on 15 and 17. Weiskopf bogeyed the first hole but then reeled off four front-nine birdies for a 31, and he added four more birdies coming back, for a 32 and 63.

Keith Fergus, Lon Hinkle, and Mark Hayes were the three with 66s.

With a birdie on the first hole of the second round, Weiskopf moved to 8 under and seemed to be loose and relaxed. But he bogeyed the second hole as he 3-putted, double-bogeyed the sixth and also 3-putted the seventh, falling to 4 under. He finished with a 75, bogeying 16, 17, and 18, and ending up with 138 for 36 holes.

Nicklaus was also struggling. He birdied the first and third holes, but 3-putted the sixth from 45 feet, bogeyed 11, and double-bogeyed 12. He made up a stroke with a 3-foot birdie putt on the monstrous 630-yard seventeenth and saved par with a delicate chip on the eighteenth hole to stand at 71 for the round and 134 for the Open.

However, Isao Aoki, the great Japanese player, moved into position to challenge Nicklaus, who was his playing partner all four days, with a round that saw him use just 23 putts in scoring a second straight 68. Aoki had just one 2-putt green on the back 9—the twelfth—and for 36 holes, had used only 50 putts. Weiskopf had used 71 putts for the same 36 holes. At the end of the second day, Aoki, with 136, shared second with Hinkle, Fergus, and Mike Reid. Next, at 137 was Hayes, followed by Weiskopf and Pat McGowan. Hale Irwin had his second 70 but was never a factor.

Nicklaus and Aoki dueled head to head on the third and fourth days. Aoki's putting slipped the third day as he used 31 puts, but his 68 to Nicklaus's 70 moved him into a share of the lead at a record 204. Twice in that round Nicklaus 3-putted, and though he reached the 540-yard finishing hole in 2, he 3-putted.

Meanwhile, Watson was making a move. After opening with a 71 and adding a 68 in the second round, Watson shot a third-round 67 and was at 206, 2 strokes behind the leaders and one behind Hinkle at 205. Hayes and Fergus were also in contention at 206.

The final day quickly settled down to a Nicklaus—Aoki head-to-head battle. Hayes had dropped out early, finishing with 74, but Fergus hung on and could have been a more serious threat had some putts fallen. Watson at one point actually passed Aoki and moved into second place, a shot behind Nicklaus, but

Presentation Ceremony, 1980

One hole to go

Jack Nicklaus

he soon fell back and finished with 71.

In the Nicklaus–Aoki "match-play" over the final 18 holes, Nicklaus used a mid-round swing correction to great advantage. Nicklaus went 2 strokes ahead with a birdie on the third hole after Aoki bogeyed the second. But his driver deserted him as he drove into the rough at the sixth and seventh holes, and he was able to save par but once as Aoki moved to within a stroke with his birdie at the eighth. But Aoki bogeyed the ninth, and Nicklaus, correcting the swing flaw, made par and after the two exchanged birdies at the tenth, Nicklaus's 2-stroke lead was intact and remained so the rest of the way.

But not without a few scares. They stood on the seventeenth tee still 2 strokes apart, and although both were on the green in 3, Nicklaus found himself looking at a 20-foot birdie putt to match what would certainly be an Aoki birdie from 5 feet. When the putt tumbled into the hole, the enormous crowd following the two players roared wildly and Nicklaus grinned broadly.

Without a miracle eagle 3 at the final hole, Aoki could not win—but he nearly came up with that eagle. His third shot, a little pitch from the bottom of the hill, missed going in by inches and stopped 3 feet away. Nicklaus played a 3-wood from the tee, a 3-iron short of the green, and a pitch 10 feet from the hole. Nicklaus was just trying to roll the putt close, but it fell in, and the crowd encircling the eighteenth green broke through restraining ropes and charged Nicklaus.

Nicklaus help up his hand, palm toward the gallery, stopping the crowd long enough for Aoki to also make his birdie putt. Then the crowd let loose again, roaring their approval, and chanting, "Jack, Jack, Jack."

At the age of 40, an age at which only three men had previously won an Open, Jack Nicklaus had won his fourth. When asked how he felt about his victory, Nicklaus replied, "You'll never know how sweet it is!" The big leaderboard near the nineteenth green had a message: "Jack is Back." It hardly seemed that he had been away.

Arnold Palmer

Tom Watson

Metropolitan and New Jersey Area Championships at Baltusrol

Men's Championships

One Metropolitan Open Championship (1908), two Metropolitan Amateurs (1908, 1912), six New Jersey State Amateurs (1903, 1907, 1913, 1925, 1941, and 1979), one New Jersey State Open (1983), one New Jersey PGA (1935), and one New Jersey Junior Championship were played at Baltusrol—twelve in all. A listing of the more important ones emphasizes the extent to which Jerome Travers of Upper Montclair Country Club dominated amateur golf in the 1907–1915 era—a period in which he also won four national amateur championships and the U.S. Open.

The great Walter Travis also appeared in the 1908 and 1912 Metropolitan Amateurs, although by 1912 he was 50 and no longer the golfer he had been when he won the U.S. Amateur in 1900, 1901, and 1903, and the British Amateur in 1904.

Baltusrol members played in several of the amateur championships selected for description here, including Max Behr (1907 New Jersey Amateur and 1912 Metropolitan Amateur), Wallace Sinclair and C. G. Sullivan (1912 Metropolitan Amateur), and Augie Kammer and E. M. Wild (1924, 1925, and 1926 New Jersey Amateurs). Actually, Kammer beat Wild three times in those New Jersey Amateurs. Only the 1925 match was played at Baltusrol; the 1924 and 1926 tournaments were held at the Shackamaxon and Deal clubs, respectively. The rivalry between the two Baltusrol greats is described in the section on Baltusrol golfers.

1907 New Jersey Amateur

Jerry Travers of Upper Montclair beat Max Behr of Baltusrol in the 1907 New Jersey Amateur. Travers also won the U.S. Amateur Championship in 1907 and would win it again in 1908, 1912, and 1913. He would also win the National Open in 1915. As said

earlier, he dominated U.S. Amateur golf in the middle teens.

Behr was club champion at Baltusrol in 1909 and New Jersey State Amateur in 1909 and 1910. He defeated Jerry Travers in the 1910 match. Behr was also a noted golf architect and was publisher of *Golf Illustrated*.

1908 Metropolitan Amateur

In the 1908 Metropolitan Amateur, Charles H. Seely beat Jerry Travers on the thirty-eighth hole, after Travers hooked his tee shot to the rough and took 5 on the hole. Seely was on the green in 3 and holed a 15-footer to win. Seely had defeated Walter Travis in the semifinals.

1912 Metropolitan Amateur

In the 1912 Metropolitan Amateur, Jerry Travers of Upper Montclair beat Oswald Kirkby of Englewood in a 36-hole final, 9 and 8. He was 9 up at the eighteenth hole, with 74. Kirkby had defeated the former great, but now aging, Walter Travis in the quarterfinals, 1 up. This was the second year Travers beat Kirkby in the final.

The tournament was of special interest because of the Baltusrol members who participated.

There was Wallace Sinclair, who won his first two matches and then defeated his fellow club member C. T. Sullivan in the quarterfinals. He was then beaten by Kirkby in the semifinals, 2 and 1.

Sullivan had beaten Max Behr, another Baltusrol member, in the first round. Sullivan was a three-time Baltusrol club champion—1907, 1908, and 1911.

1935 New Jersey PGA

Maurrie O'Connor of Branch Brook won the 1935 New Jersey PGA Championship by defeating Byron Nelson of Ridgewood Golf Club. They were tied after 72 holes at 291. They played 18 holes the following day and were still tied at 75. O'Connor won the last 18 holes with a 71 to Nelson's 73.

One hundred eight holes was the longest state PGA. Byron Nelson was in his first year as assistant pro to George Jacobus, then PGA president. Jacobus told Nelson to practice—which he did. For Nelson subsequently won the U.S. Open, PGA, and Masters among his 54 Tour titles, which included 11 consecutive victories in 1945.

Women's Championships

Three Women's Metropolitan Championships—1903, 1964, and 1983—were played at Baltusrol.

Also five Women's New Jersey Championships were played there—1949, 1953, 1958, 1973, and 1983.

Baltusrol Golfers

Men Golfers

Just as Baltusrol itself has been in the mainstream of American golf since 1895, so have its golfers. Their names will be found on the lists of club champions or seniors association champions; and from there they ventured forth to distinguish themselves by winning outside tournaments as well, including the New Jersey Open, Amateur, and Seniors Championships; the U.S. Amateur and Seniors Championships; and the International Seniors.

There have been 179 of these men's events between 1895 and 1984: 86 club and 63 senior association championships (the seniors association was not formed until 1922) and 30 outside championships. One hundred forty-three of these events, or 80 percent, were accounted for by the 30 individuals who won more than one event. Indeed, 14 of these men won 104 events or 58 percent of the total. They were:

Martin S. Lindgrove 19
R. C. James 11
A. R. Kammer 10
E. M. Wild 9
Lawrence E. Carpenter, Jr. 9
Richard A. Henry 7
John A. Dietrich 6
Robert Finney 6
Max Marston 6
John Poinier 5
John Farrell, Jr. 4
Thorwald Larson 4
Homer Lichtenwalter, Jr. 4
James A. Tyng 4

Of course, golf championships vary in their importance. The New Jersey Open is more prestigious than the Amateur; and a club championship rates higher than a seniors championship. But that does not affect the relative standing of these players significantly, since most competed successfully at several levels.

E. M. Wild

James A. Tyng et al. (1896–1915)

James A. Tyng was an example. He and Louis P. Bayard were club-champions-to-be at Baltusrol. They participated in the second USGA Amateur tournament held at Shinnecock Hills in July 1896. They were joined by fellow member Henry P. Toler.

Although Tyng played from the Morris County Golf Club as well as from Baltusrol, he won the Baltusrol club championship in 1899 and 1900. He also won the U.S. Seniors Championship in 1912 and 1915. According to *Fifty Years of American Golf*, Tyng made baseball history as well as golf history, having been both a catcher and a pitcher while at Harvard in 1876. He is said to have been the first catcher to don a catcher's mask.

Bayard, who won the club championship in 1904, served the longest term as club president of Baltusrol—from 1898 to 1919.

Toler, a broker, later distinguished himself as the hero of the 1909 Baltusrol clubhouse fire by "swinging the women out the windows of the burning building."

All three, Tyng, Bayard, and Toler, qualified for the 1896 Shinnecock Hills tournament, but they were eliminated before the semifinals. The tournament winner was H. G. Wingham of Chicago, a son-in-law of the defending champion, Charles B. Macdonald.

Then, in September 1896, Tyng and Toler were red-hot favorites to win the final at an amateur foursome competition staged at the Newport Golf Club. But they were beaten by Fred Havemeyer, age 16, and Harry Havemeyer, age 18, sons of "Sugar King" Theodore Havemeyer, who was the first president of the U.S. Golf Association.

Max Marston (1914–1923)

Max Marston was a contemporary and worthy competitor of Bobby Jones, George Von Elm, and Chick Evans in the middle 1920s. He had won the Baltusrol club championship in 1914, 1915, and 1916, and the New Jersey Amateur in 1915 and 1919. He resigned from Baltusrol in 1918 to move to Philadelphia, and there won three Pennsylvania Amateurs. Marston defeated Francis Ouimet in the semifinals of the 1923 U.S. Amateur Championship, and then beat the outstanding Jess W. Sweetser to win that tournament. Sweetser had beaten Chick Evans to win the Amateur the preceding year. Both Marston and Sweetser were members of the 1922 U.S. Walker Cup team.

According to *Who's Who in Golf*, Marston, like Mac Smith, "never took a divot."

On October 11, 1923, Marston was made a lifetime honorary member of Baltusrol.

E. M. Wild and Augie Kammer (1920s)

E. M. Wild and Augie Kammer were outstanding rivals at Baltusrol for years.

Wild lived in Cranford. He was with a wire rope company and later in life insurance. He was of slight build, weighing only 135 pounds. And he was handicapped by a left arm that had been injured in an iceboating accident. Nevertheless, he won the Baltusrol club championship eight times in a row, 1922–1929, and again in 1931. Wild's daughter, Gail, was a chip off the old block. She won the New Jersey Women's Championship in 1937 and the Baltusrol women's club championship in 1939, 1940, and 1942.

Kammer lived in South Orange and was an executive with a glass company. He had won the Baltusrol club championship twice–in 1919 and 1921. And he won the New Jersey Amateur in 1924, 1925, and 1926. He was noted for his excellent putting. An Augie Kammer putting contest is held annually at Baltusrol on Labor Day weekend. Kammer's son, Augie Jr., who was associated with the Seminole Essex Wire Corporation, a Ford Motor supplier, was also a chip off the old block. He won the Baltusrol club championship four times—in 1930, 1936, 1937, and 1938.

Somehow Wild could not beat Kammer in an outside tournament—notably in the New Jersey Amateurs of 1924, 1925, and 1926.

1924 New Jersey Amateur. In the 1924 New Jersey Amateur, which was played at Shackamaxon Golf and Country Club, Kammer beat Wild, 1 up. Wild had beaten Fred Connell of Glen Ridge, 3 and 2, in the semifinals. And Kammer had bested Tom Green of Princeton, 1 up, on the nineteenth.

In the final, Kammer and Wild were even at the turn, but Kammer won 10, 11, and 13, to go 3 up. Wild won 14 and 15, and they halved 17 and 18.

Wild's bad luck continued to plague him—he tied for the medal with two others but lost on the coin toss.

1925 New Jersey Amateur. With an 80, Kammer had trouble qualifying for this tournament, which was held at Baltusrol. He would have been one of six playing off for five places, but one withdrew.

Kammer beat Jack Sharkey, a medalist with 71, in the second round with a great rally. Kammer was 1 down after 15 but rallied with birdies on 16, 17, and 18, to win 2 up. He defeated Wild, 3 and 2, in the semifinals and then beat Bill Reekie of Upper Montclair in the final, 2 and 1.

1926 New Jersey Amateur. In the 1926 New Jersey Amateur, which was played at Deal Golf and Country Club, Kammer again beat Wild, 5 and 4, in the final after defeating Reekie in the semifinals. This was the third straight victory for Kammer. "Deadly putting was Kammer's forte," *The New York Times* said, "and today he was almost uncanny."

Unfortunately, Wild reached the semifinals and lost in five New Jersey Amateurs.

R. C. James (1920s)

The Baltusrol seniors tournament was virtually dominated by R. C. James. He won the championship in ten successive years (1923–1932).

According to member Phil Hartung, James had been employed by United Gas Improvement Company in Philadelphia before moving to New Jersey to join the Public Service Company in Newark. Actually, he moved right into the heart of the Baltusrol golf course itself. He occupied the house that was formerly located in the triangle at the right of the first, second, and third Lower fairways. There were a few such houses on the land that Louis Keller had acquired in the 1890s, and the club apparently rented them.

Although the house has long since been razed, its water well, which is said to be the best on the Baltusrol property, is still operating.

According to Maitre d' Ernie Mudd's wife Rose, James was well liked by her father and by the others who worked on the golf course at the time: "He made everyone feel so important."

James was made an honorary member in March 1960.

Martin S. Lindgrove (1927–1960s)

Born in 1889 in Kansas, Martin S. Lindgrove was a U.S. Air Force pilot from 1917 to 1922 (1st lieutenant), an investment banker until 1935, and a retiree until World War II, when he reentered the air force as a major and served as senior port officer in New York City and San Francisco. He retired as a full colonel.

During his ten years in the army, Colonel Lindgrove played a great deal of golf at Garden City Golf Club and other Long Island clubs and at the Presidio Golf Club in San Francisco.

Colonel Lindgrove joined Baltusrol in 1927 and was club champion four times (in 1934, 1935, 1939, and 1940) and seniors champion six times (in 1940, 1941, 1946, 1951, 1952, and 1953).

He also won the New Jersey Seniors Championship eight times—in 1947–1950, 1952–1954, and 1957—and the U.S. Seniors Championship in 1947 along with the International Seniors Championship in 1951.

Lindgrove was also captain of the U.S. Seniors Golf teams to Great Britain and Europe in 1954 and to Africa in 1955.

A kindly man and a pleasant golf companion, Colonel Lindgrove competed with three other Baltusrol members in the 1946 U.S. Seniors Tournaments—A. F. Kammer and Charles P. and Newton A. Burgess. Lindgrove, Kammer, and Charles Burgess all scored 155 for the first 36 holes but lost to Ellis Knowles in the 18-hole playoff.

Robert Finney (1930s–1960s)

As stated earlier, Bob Finney joined Baltusrol immediately after graduating from Yale in 1923, where he had pitched on the baseball team. His father had been a Baltusrol member from 1915, so as a boy Bob not only played the Old Course, he watched the construction of the new ones from 1918 to 1922.

Bob Finney won the club championship in 1933 and the seniors championship five times—1960, 1961, 1964, 1965, and 1968.

He served on most of the committees of the board of governors beginning in 1926—although he was not elected to the board itself until 1947 and he served for nine years. He was reelected to the board in 1967 after having been chairman of the 1967 U.S. Open and he served until 1976, when he was made governor emeritus. He was club president from 1967 to 1970.

Bob Finney was the top executive of Street & Finney, an advertising firm. Because of his many years of service with the club, his knowledge of the club and golf generally, and his unusual leadership

Colonel Martin Lindgrove

Homer O. Lichtenwalter, Jr.

abilities, Finney ultimately became known as "Mr. Baltusrol."

Homer O. Lichtenwalter, Jr. (1950s–1960s)

Homer Lichtenwalter, born in 1920, graduated from the University of Florida with degrees in chemistry and chemical engineering. He was president of a family-owned manufacturing company and later vice president of Presbyterian Homes of New Jersey. He won the club championship at Canoe Brook Country Club in 1940 and was a four-time champion at Baltusrol—in 1950, 1957, 1961, and 1962.

Richard A. Henry

Lichtenwalter played in the U.S. Amateur Championship in 1952 and 1953 and was a medalist in the New York sectional qualifier in 1953.

He is a golf statesman, serving on Baltusrol's board of governors and as chairman of its golf committee from 1967 to 1973. And he was president of the New Jersey State Golf Association from 1981 to 1983.

Richard A. Henry (1940s–1960s)

Dick Henry was born in Concord, New Hampshire in 1903. He graduated from Dartmouth in 1924 and from Tuck School in 1925. Dick was a certified public accountant and an authority on public utility accounting. He was a managing partner of Niles & Niles in New York and later a partner in Deloitte, Haskins & Sells, into which his firm had merged.

Henry won the Baltusrol club championship in 1944, 1946, and 1955 and followed those victories with the Baltusrol seniors championship in 1956, 1962, and 1967. He was a New Jersey Seniors champion in 1969.

Dick served on the club's board of governors as treasurer.

Dave Baldwin and Larry Carpenter (1940s–1970s)

Golf's durability as a lifetime sport and its role in nurturing lifelong friendships is well known. It is a shared experience that binds and enriches the lives it touches. No doubt it has been the principal catalyst in the 40-year friendship of David M. Baldwin and Lawrence E. Carpenter, Jr.

They were both born in New Jersey in 1928. They met at junior golf tournaments when they were 15 or 16 and became close friends.

They were all-around athletes in their respective prep schools. Dave was a football star, runner, shot putter, and golfer at Pingry. Larry had letters in football, baseball, basketball, and golf at Hotchkiss and was designated outstanding athlete of the 1947 senior class.

Both graduated from Lehigh University in 1951 with degrees in finance. And both were scratch golfers and co-captains of the golf team. Larry was also pitcher on the baseball team.

Larry Carpenter's father had joined Baltusrol in 1937; Dave's father joined in 1959 after belonging to the Essex County and Rock Spring clubs.

The Carpenters, father and son, defeated the Baldwins, father and son, to win the New Jersey Father–Son Championship in 1948.

Dave won the club championships at Bridgehampton in 1945; Essex County in 1948, Rock Spring in 1951 and 1953, and Baltusrol in 1963.

Larry won the club championship at Baltusrol eight times between 1948 and 1973—a remarkable span of 25 years. Dave was runner-up or medalist in three of those. Carpenter also won the club's senior association championship in 1984.

Dave won the New Jersey Open Championship in 1954 and was runner-up in the National Left Hander's Tournament, in which he won the long-driving contest.

Meanwhile, that same year, Larry reached the quarter-finals in the British Amateur Championship and the semi-finals in the French Amateur, and he won the German Open Championship in Nuremberg.

In those years Dave and Larry also teamed up to win the New Jersey Scotch Foursome Championship three times.

Each has four children and works in New York. Dave is president of a major real estate firm, and Larry is a Wall Street investment banker. And each still roots for the other in everything—except when they are competing in the same golf tournament. There, understandably enough, they are arch rivals.

William Y. Dear, Jr. (1950s–1970s)

Billy Dear has played championship golf for over 40 years despite 42 surgical operations, including a complete shoulder replacement, 2 chrome knees, 4 fused vertebrae, and 5 plastic fingers.

He was New Jersey State Amateur champion in 1951 and runner-up in 1941, 1946, 1953, and 1957.

He was runner-up in the New Jersey Open in 1951.

And he was Baltusrol seniors champion in 1969 and 1970 as well as New Jersey Seniors champion in 1976. Also, he won the Metropolitan Seniors Championship in 1971.

Born in 1912 and a graduate of Brown University, Dear's first golf championship was the New Jersey Interscholastic; he later won the National Interscholastic as well.

David M. Baldwin and Lawrence E. Carpenter, Jr.

John Poinier

William Dear, Jr.

A former president and tournament chairman of the Metropolitan Golf Association and active in other golf associations, this determined individual also took time out to operate the Jersey City Printing Company.

John Poinier (1950s–1970s)

A football, baseball, and hockey player in prep school, Jack Poinier graduated from Yale in 1934. In 1985 he was a retired chartered life underwriter and insurance company executive.

Poinier won the club championship at Canoe Brook in 1953 and at Baltusrol in 1965. He also won the seniors championships at Baltusrol in 1972 and 1973, the New Jersey State Golf Association Seniors Championship 1971, and the New Jersey State Seniors Golf Association Championship in 1970.

He was president of both the New Jersey and the Eastern Seniors Golf Associations and was a member of the board of governors at Baltusrol.

Thorwald Larson (1960s–1970s)

Thorwald "Tom" Larson was born in Varborg, Sweden, in 1907. He graduated from MIT in 1928 with a degree in chemistry and spent most of his career with Carbic Color & Chemical Company, now American Hoechst.

Before joining Baltusrol, Tom won club championships at several clubs, including Rock Hill (S.C.) Country Club (1941), Myers Park Country Club, Charlotte, North Carolina (1945), and Rhode Island Country Club, where he was club champion several times in the years 1947–1959.

By the time he joined Baltusrol, Tom was a senior, and he promptly went about winning the club's seniors championship in 1966, 1974, and 1978 as well as the New Jersey Seniors Championship in 1967.

John A. Deitrich (1960s–1980s)

John Deitrich, born in 1912, earned a B.S. from Muhlenberg College, a MS from Lehigh University, and an MBA from Harvard Business School. He also earned five varsity letters in football and baseball as an undergraduate and was a teacher and coach at Allentown Prep. He was a late bloomer in golf, winning his first championship, the Bermuda Invitational, in

1962, and the Breakers Seniors in 1970.

John won the senior championship at Baltusrol six times—in 1975, 1977, 1979, 1980, 1982 and 1983.

In 1985 Deitrich had been engaged in the steel business for almost 40 years in a principal capacity. And he has generously contributed his considerable talents to the advancement of education and health care.

John Farrell, Jr. (1960s–1980s)

John Farrell, Jr., worthy son of a celebrated father, graduated from St. Benedict's Prep in 1951 and from Georgetown University in 1955. Winner of the New Jersey Junior Championship in 1950 and the Korean Air Force Championship in 1956, Farrell was also a four-time club champion at Baltusrol—in 1964, 1969, 1972, and 1981. And he was runner-up in two of the intervening years. He also won the Augie Kammer putting championship at Baltusrol three times.

John Farrell, a former member of Baltusrol's board of governors, became a stock broker in 1968.

John Farrell, Jr.

John A. Dietrich

Women Golfers

arly women golfers at Baltusrol were as active and as enthusiastic as the men—perhaps more so. And with their strong penchant for order, they began by structuring the sport.

In 1898 they participated in the founding of the Women's Metropolitan Golf Association, which is one of the oldest and most prestigious women's golf organizations in the United States. The WMGA now has some 1,850 members from more than 150 clubs in New Jersey, Long Island, and Westchester County. Membership is open to amateur women golfers with handicaps of 20 or less. Baltusrol is one of only six clubs that have maintained uninterrupted membership in the WMGA.

Baltusrol member Mrs. W. Fellowes Morgan was WMGA's first president, serving from 1899 to 1903. Mrs. Morgan was also an early member of the Morris County Golf Club, which, you will recall, women organized in 1894, although they were deposed by the men less than a year later.

Another Baltusrol member, Mrs. Edward A. Manice, succeeded Mrs. Morgan as president of WMGA and served for one year (1904).

Five other Baltusrol women also served as presidents of the WMGA. Mrs. Myra D. Patterson was one (1910–1914). She was president during several other periods as well but as a member of other golf clubs. Other presidents from Baltusrol were: Mrs. Frederick B. Ryan (1930–1933), Mrs. Wright D. Goss, Jr. (1942–1945), Mrs. William Hockenjos, Jr. (1948–1950), and Mrs. Theordore Hawes (1955–1956).

Three Baltusrol women participated in the first WMGA tournament, which was held in June 1900 at the Morris County Golf Club. They were: Mrs. Morgan and Miss M. Oliver and Miss L. DeBary. That tournament was won by 17-year-old Genevieve Hecker, who defeated Mrs. Ruth Underhill. Mrs. Underhill was the first secretary of the WMGA and won the U.S. Women's Championship in 1899. Miss Hecker, who later became Mrs. Charles Stout, won the WMGA championship three more times—in 1901, 1905, and 1906. She also won the 1901 U.S.

Women's Amateur at Baltusrol and then successfully defended her title the following year at The Country Club in Brookline, Massachusetts.

Some thought Miss Hecker the equal of any woman golfer in the world. She is remembered equally for her writings. She did a series of articles for *Harper's Magazine* in 1902. And her book *Golf for Women* was literally the golf primer of her time—it helped carve a prominent place for women in golf. Miss Hecker is quoted as saying that women golfers are pluckier than men—they "show a relatively greater degree of nerve particularly on the putting green."

Here's an excerpt from Genevieve Hecker's book:

> All golf-balls should be kept for some time before being used . . . the gutty ball from eight to nine months to a year, and rubber-filled a little less. After the ball is used, the most satisfactory thing to do and most economical as well, is to send it to a professional, or to one of the big ball-making manufactories and have it remade. Any ball manufactory will do it for one dollar to two dollars a dozen.

Obviously, those early Baltusrol women played in pretty fast company. Indeed, Mrs. E. A. Manice from Baltusrol, the second president of the WMGA, won the WMGA Match Play Championships in 1902, 1903 and 1904. Mrs. Manice was mentioned frequently in early golf books. Like Miss Hecker, she was one of the few women of the day "who used a full backswing and a recognizable pivot." She won the driving competition at Ardsley in 1898 but lost in the 1900 contest when her 189-yard drive was exceeded by 4 inches. That was considerable in the days of gutta percha balls.

And Mrs. Myra Patterson, who served many terms as president of WMGA and was an early club champion at Baltusrol, was runner-up in the 1904 WMGA championship. She was also the first winner of the North and South at Pinehurst in 1903 and was winner again in 1904 and 1906. And she qualified for the 1905 National Championship.

It would be many years before the WMGA championship would again be won by Baltusrol women. But they returned to the top when Mrs. A. Sherburne Hart ("Hatsy") won in 1970, and Miss Lee Steffens triumphed in 1984. More about these women later.

Women's New Jersey Golf Association

Baltusrol furnished ten presidents of the Women's New Jersey Golf Association, four of whom had also headed up the Women's Metropolitan Golf Association. They were:

Mrs. Frederick B. Ryan	1925–1926
Mrs. Clifford Armitage	1939–1940
Mrs. Donald Baldwin	1947–1948
Mrs. Theodore W. Hawes	1949–1950
Mrs. Homer Lichtenwalter	1953–1954
Mrs. George A. Murray	1957–1958
Mrs. Reginald T. Blauvelt, Jr.	1963–1964
Mrs. Frank Stackhouse	1971–1972
Mrs. Kenneth Wood	1973–1974
Mrs. John Swart	1983–1984

Also, Baltusrol women won ten New Jersey State championships. Helen Hockenjos won six of these, Mrs. A. Sherburne Hart two, and Miss Gail Wild and Mrs. Wright Goss, Jr., one each. Gail was the daughter of E. M. Wild, a nine-time club champion of Baltusrol.

Other Women's New Jersey Golf Association Trophies won by Baltusrol women included the Uebelacker Memorial Trophy, which was won by Helen Hawes in 1957; by Winifred Baldwin in 1958, 1963, 1966, 1971, 1972, and 1977; by Pearl Black in 1961 and 1964; and by Ruth Swart in 1983.

Also the Nelson Memorial Trophy, which was won by Winifred Baldwin in 1951, by Bette Gaertner in 1968, by Brita Blauvelt in 1975, by Nancy Rogers in 1982, and by Mary Stackhouse in 1983.

And finally, the Dickinson Trophy, which was captured by Winifred Baldwin in 1960, by Mary Stackhouse in 1969, by Ruth Swart in 1973, and Mary Stackhouse in 1983.

Garden State Women's Golf Association

The need for keener competition and smaller fields that could move faster became clear to the metropolitan women tournament players after World War II. This led to the formation of three sectional women's golf associations for 12-and-under (now 14-and-under) handicappers. They were the Women's Cross County in Long Island (1949), the Women's Tri-County in Westchester (1950), and the Garden State Women's Golf Association (1953).

Helen Hockenjos was one of the 11 women present at Garden State's organizational meeting. She was elected its first president, and she won its first tournament in 1954.

Women Club Champions

The records on women club championships at Baltusrol are not as complete as the records on the men. Although several women have been identified as early club champions, the first woman club champion listing in the record is for the year 1916, and there are several gaps in the record between that year and 1929, when the record firmed up.

As with the men's golf, a relatively few women won most of the 60 club championships of record. Actually, 13 women won 46 or 76 percent of these events, and 7 of these women won 34 events or 56 percent. They were:

Helen Hockenjos	13
Charlotte DeCozen	5
Collette Ramsey	4
Elizabeth A. Hardin	3
Isabel Mercer	3
Alice Rutherford	3
Gail Wild	3

The Baltusrol club championship, of course, is but one measure of the relative proficiency of the women golfers.

Helen Hockenjos, for example also won six New Jersey championships, and Gail Wild won one.

Also Mrs. A. Sherburne Hart won the New Jersey Women's Amateur twice.

And Helen Hawes, a two-time Baltusrol club

champion, won the WMGA Seniors Championship
and the prestigious U.S. Senior Women's Golf Association Championship.

Finally, Lee Steffens, 2-handicapper, won both
the Women's Metropolitan Golf Association Championship and the Garden State Championship in June
1984. Two months later she defeated Robin Jervey to
win the 1984 Baltusrol club championship. Robin, a
student at Furman University and a 5-handicapper,
had won the club championship in 1982 (at age 18) and
again in 1983. More will be heard from these young
women.

Helen Hawes (1940s–1960s)

Helen Hawes (Mrs. Theodore W. Hawes) was born in
Illinois and was educated at Eureka College and
Northwestern University. She won seven club championships at Canoe Brook Country Club before joining Baltusrol, where she won the club championship
twice. She also won the WMGA Seniors Championship in 1959 and the U.S. Senior Women's Championship in 1961, having been runner-up in the U.S.
Seniors in the preceding three years.

In addition to her presidencies of the WMGA
and the New Jersey Golf Association, Helen Hawes
was a member of the prestigious Women's Committee
of the United States Golf Association for several years
and served as its chairperson in 1964–1967. In 1967
she was nonplaying captain of the U.S. Curtis Cup
Team.

Helen Hockenjos (1930s–1960s)

Helen Hockenjos had more than a taste of the big
time in her 38 active years of golf.

Glenna Collett Vare dominated women's golf at
the time that Helen first took up the sport in 1929.
Glenna was six-time U.S. champion and two-time
Canadian champion. She was considered the greatest
American woman amateur of the 1920s and 1930s,
with an overall record of 49 championships.

And Maureen Orcutt was only a cut below. She
won the Canadian Championship twice, the North
and South three times, the Eastern seven times, and
the WMGA ten times. Her WMGA victories spanned
42 years—from 1926 to 1968.

Helen Hockenjos defeated Maureen Orcutt to
win her first New Jersey State Championship in 1936.

Helen Hawes

Lee Steffens

Robin Jervey

Harriet Hart

Collette Ramsey

Helen Hockenjos

And she defeated Glenna Collett Vare in 1940 in the final of the Berthellyn Cup International at Huntingdon Valley, Pennsylvania, after beating Philadelphia's popular Helen Sigel in the semifinals.

In all, Helen Hockenjos won 6 New Jersey State Championships and a record 13 Baltusrol club championships, the latter in the 20 years from 1948 to 1968.

Helen won many other tournaments as well, including Fred Waring's Shawnee Women's Invitational (1937) and New Jersey Pro-Lady Match Play (1948, 1949, 1950, 1951). And as New Jersey champion, Helen competed for New Jersey in most intersectional team and intercity matches.

In 1983 she was named to honorary membership in both the Garden State Women's Golf Association and the Women's New Jersey Golf Association.

Collette Ramsey (1940s–1950s)

Collette Ramsey, wife of Hobart C. Ramsey, who was president of Baltusrol in the early fifties, was one of the fastest-rising stars in New Jersey women's golf. Encouraged by her husband, who was a 6-handicapper, Collette first took lessons and incidentally developed a classic golf swing in the early 1940s, won four successive Baltusrol club championships in the middle forties (1943–1946), and was playing big-time golf by the late 1940s, when she played against Babe Didrikson Zaharias, Patty Berg, Betty Jameson, and Maureen Orcutt among others in the 1949 Women's Eastern Open Championship at Essex Fells Country Club. The Babe, however, won that one going away.

Although generally limited to one- or two-day competitions because of a chronically bad back, Collette participated in and won one five-day tournament, the New Jersey Invitational held at North Jersey Country Club in 1947.

And despite major surgery in the middle fifties, she teamed with Caroline Cudone and Willie Turnesa to win several Metropolitan Best Ball Championships.

Collette's major interest, in addition to her two daughters, is the Deafness Research Foundation, which she founded in 1955. With an annual budget approaching $1 million, the foundation provides funds to 60 leading universities and research centers for ear and hearing research.

Collette Ramsey donated the much-coveted Ramsey Tournament Trophy at Baltusrol.

Harriet Hart (1960s–1970s)

Married to A. Sherburne Hart, lawyer-executive of Union Carbide, and mother of four, Harriet ("Hatsy") Hart took up golf at age 24 between child two and child 3. She literally taught herself by reading golf instruction books and hitting whiffle balls into the back of their house. She played her first 18 at a public course in Florham Park, New Jersey, finally getting the ball airborne only at No. 18, but that was enough to bring her back.

The 1960s were tournament years for Hatsy, culminating in her New Jersey championships in 1969 and 1970 and her 1970 Women's Metropolitan Golf Championship. She won the club championship at Baltusrol in 1973 and reached the quarterfinals in the U.S. Women's Amateur when it was played at Montclair in 1974. Then with her four children in their teens, Hatsy qualified for the LPGA Tour at age 40—no doubt the oldest rookie in pro sports. In two years she played in ten tournaments, mostly close to home, making a number of cuts and having several respectable showings.

Hatsy was reinstated as an amateur in 1979 mainly due to a change of LPGA rules that required more tournament appearances than she could handle with her maturing family.

If women bring an additional dimension to the game of golf, it surely is their exquisite appreciation of the natural surroundings in which the game is played and their ability to describe them. This was beautifully done by Hatsy in a letter to Nancy Rogers in June 1984, describing a magic moment in Baltusrol's Magic Kingdom—the Upper golf course. (Nancy was women's club champion in 1976 and chairman of women's golf in 1977–1978).

> Late one still summer afternoon my husband and I were carrying a few clubs for a leisurely 6 holes. The sun was gradually sliding behind the hill and the temperature was ideal but the outstanding feature of the arriving evening was its tranquility. We were in a fairyland of light and shadow, warmth and stillness. Best of all was that every creature that afternoon seemed to feel a kinship in nature. On the second hole 2 fox kits scampered after our balls on the green, whisked back up the hill and disappeared into the woods with them. We were amused and then amazed to see the kits waiting for us on the edge of the third hole. They seemed to like golf too and knew enough to expect us at the next hole. Unfortunately

like most mothers Mrs. Fox was too dour and disapproving of the children sporting in the wrong company and so she cut off their game with us.

We were picked up at this point by a thrush. Usually shy birds, this individual just stood on the green about 12 feet from the cup. After I hit my ball and approached the green I expected a flutter and flurry of escaping wings. Not at all—I was permitted to putt and remove my ball from the cup without a notice from the thrush just a few feet from me.

Bewitched, we proceeded to the 4th tee where the view toward the setting sun silhouetted a large doe. She stood regally surveying us not 25 yards down the fairway until she casually chose her retreat into the woods.

The playback to the clubhouse was uneventful but this 1985 USGA Women's Open course really had become that afternoon a magic kingdom. I have had other golf experiences like this one of pure nature enchantment and I think they add a unique aspect to the game not found in other sports. Golf can be enjoyed on so many levels.

Baltusrol Women's Golf Committee

Today there are some 330 women with links privileges at Baltusrol—almost twice as many as 30 years ago. The women's golf committee, with its four officers and nine subcommittees, oversees an active program consisting of regular Tuesday competitions for 60 to 85 eighteen-holers and Wednesday competitions for 20 to 40 nine-holers. The nine-holers organized in 1958.

Major tournaments, some of which are two-day events, are interspersed throughout the season. These include the (Collette) Ramsey Handicap, Grandmothers' Tournament, President's Cup, Anne O'Mara Silver Cup, Blauvelt Cup, Farrell Tournament, and of course, the club championship.

The women's committee also sponsors the women's golf association tournaments that are held periodically at Baltusrol, including the Women's Metropolitan, the New Jersey State, and the Garden State Championships.

But more than this, the women's committee is the only formal organization of women at Baltusrol. It is the principal representative or advocate of women golfers' interests and needs before the board of governors, including such matters as the adequacy of the locker room, the need for toilet facilities on the golf course, and the desirability of having an informal snack place where wives can join their husbands after a round of golf. The spokesperson in these cases is the women's golf committee chairperson, who invariably is a woman of substance and standing—a leader.

CLUB
PROFESSIONALS

Baltusrol has had nine professionals in its 90 years. But two accounted for 60 of those years—George Low (22) and Johnny Farrell (38).

In the earliest days of American golf, it was customary to import professional teachers from Scotland, and it was not uncommon for the pro to act as greenkeeper as well in addition to making golf clubs for sale to club members.

Tom Gourley was Baltusrol's first pro. He was hired in 1897 at $30 a month but stayed for only a year. Gourley had played in the second U.S. Open, held at Shinnecock Hills in 1896.

Brown and Low

Willie Anderson

Willie Anderson replaced Gourley in 1898 but resigned in 1899 to play tournament golf. Anderson was born in North Berwick, Scotland, where his father was greenkeeper for the North Bergen links. He emigrated to the United States in 1895 and reportedly worked at as many as ten American golf clubs over the next ten years.

As said earlier, Anderson won a record four U.S. Open Championships—1901, 1903, 1904, and 1905. The 1903 championship was played at Baltusrol. And he won four Western Opens, which then ranked only a little behind the U.S. Open.

Anderson died in 1910 at the age of 32.

Anderson was followed by David Hunter as Baltusrol's next pro. He remained for about four years, when George Low was hired to replace him.

George Low

George Low, a Scot from Carnoustie followed Alex Smith of the famous Smith family of golfers, to America in 1899. He was a top golfer, having tied for second place behind Alex's brother, Willie Smith, in the 1899 U.S. Open. He won the Metropolitan Open at Hollywood in 1908, with Alex Smith second. And he had other important finishes.

Low, a stocky fellow of medium height with a strong Scottish burr, played many exhibitions and was considered an excellent golf teacher. Presidents Taft and Harding were reportedly among his pupils.

He was also a golf course architect and helped design the Westchester Country Club course. His course design activities apparently conflicted with his teaching role, for he talked of resigning from Baltusrol several times, only to decide to remain after being granted a salary increase or a leave of absence to give him time for his New York architectural shop. In 1922 he was made an honorary lifetime member of the club.

Major Jones was hired in 1923 to replace Low in his greenkeeper role, but Low continued as the golf professional until late 1925.

Jack Forrester was hired in 1926 to replace Low as golf professional, and he was subsequently replaced by Pete Kosky.

In 1934 Johnny Farrell came to Baltusrol from Quaker Ridge in Westchester, and he remained as club pro until he retired in 1972.

Scotch singer Harry Lauder and George Low

The Legendary Johnny Farrell

Johnny Farrell, who was born in White Plains, New York, in 1901, was the U.S. Open champion in 1928. He won the title the hardest way imaginable, by beating the greatest golfer of his day (many call him the greatest golfer of all time)—Bobby Jones. He beat Jones in a 36-hole playoff after they had tied for the title at 294 at Chicago's famous Olympia Fields Country Club.

That year, the USGA decided that a playoff would be held over 36 holes, which increased the odds against Farrell's winning. Although he already had a reputation for being a fine golfer, Farrell was not ranked with a Bobby Jones in those years. But he had won several important events, and his reputation was constantly growing.

Johnny was only 21 when he captured his first tournament, the Shawnee-on-Delaware Open in 1922. He was only 19 when he played in his first U.S. Open in 1920 at Inverness Club in Toledo, Ohio, where Ted Ray beat out even his fellow English star, Harry Vardon, to become the last foreign winner of the American Open until Gary Player won at Bellerive in 1965. Playing in such company, this slim, black-haired golfer was good enough to qualify for the last two rounds of play.

After the tournament, Johnny asked the great Vardon if he could suggest anything that would make him a better player. Vardon said, "Certainly," and went into some depth, telling Johnny what he should and should not do. Even then Johnny was anxious to reach the top.

Until he finally crashed through to win in 1928, Farrell was constantly coming close in the U.S. Open. He finished fifth in 1923, and was third, a stroke behind Jones and Willie Macfarlane, in 1925. He was third again in 1926.

Then, in 1927, Farrell won eight straight tournaments, establishing a record that went unchallenged until Byron Nelson captured 11 straight in 1945. In all, Farrell captured 10 tournaments in 1927.

This is the man who came to the 1928 U.S. Open Championship as one of the many who were out to stop Jones from his continuing march to victory. Jones had already won the U.S. Opens of 1923 and 1926. He had lost to Willie Macfarlane after a double playoff in 1925. He had finished second in the 1922 Open, had won the U.S. Amateur Championships in 1924, 1925, and 1927 and the British Open in 1926 and 1927. There was no question of Jones' position as No. 1 golfer in the world.

At Olympia Fields in 1928, Farrell found himself overshadowed in the unenviable pairing with Jones:

> Bobby, of course, was the big attraction. I was scrambling. He'd hit the green in 2, and I'd still be off. He would putt out, and I would still have a putt, but he would start leaving for the next tee, and there I was, trying to hole out. It was a little disconcerting. I shot a 77.
>
> So I asked Bobby if he would mind marking his ball until I finished putting. Someone else might have refused. But you know what sort of sportsman Jones was.

In the playoff for the 1928 Open, it was Farrell who survived. He had led the morning round with a 70 to a 73. Then he fell behind in the afternoon and was down a stroke to Jones going to the thirty-fifth hole. Few thought that Johnny could come back so near to the end, after having blown a 3-stroke lead. But he did, with birdies on the last two holes, to prove once again that behind the gracious smile was a fighter and one of the finest putters in the game.

Johnny Farrell virtually dropped out of competitive golf after taking the Baltusrol job in 1934, although he won the New Jersey Open in 1936. But he remained an active part of golf. There are many firsts to Farrell's credit. In addition to being the first "best dresser" among the playing brethren, he made many appearances at sports shows, and he performed in the vaudeville days of the Hippodrome and the Palace in New York. He made golf movies and produced one of the first TV golf shows. And he became a member of the Golf Hall of Fame in 1961.

And the Farrell name lives on competitively.

In 1985 Johnny's sons—Billy, Jimmy, and Johnny Jr.—were all fine golfers. Billy was a well-known professional with the New Jersey Open title among his many triumphs. Johnny Jr. was a four-time club champion at Baltusrol and a member of the board of governors. And Kay, Johnny's charming wife, was still playing competitively in women's golf and continuing her long-time activities in promoting charities through golf. The Farrells' two daughters are Mrs.

Adapted from "Johnny Farrell—A Baltusrol Institution," by David Isenberg, and "Johnny Farrell's 60 Years in Golf," By Gary Dorfman, which appeared in the 1967 and 1980 USGA Open programs, respectively, published by Baltusrol Golf Club; permission from Mr. Dorfman.

Peggy McGuire, who is married to a member of the club, and Mrs. Cathy Rock, of Westfield, New Jersey. The Farrells were recognized as the Golf Family of 1966 by the Metropolitan Golf Writers Association, and they received the Bill Chapin Memorial Award at a National Awards Golf Dinner.

When Johnny retired from Baltusrol, the members dedicated a large room on the second floor of the stately clubhouse to the remembrance of his long service to Baltusrol and professional golf. The club also made him an honorary member.

A handsome color photograph of Johnny dominates the room. It is flanked by letters of tribute from then President Richard M. Nixon, a former member, and the late Duke of Windsor, who had been a frequent visitor to Baltusrol and a friend of the Farrells for 40 years.

When last spoken to, Johnny was still playing and teaching from his course-side home at the Country Club of Florida in Delray Beach, Florida.

David P. Carolan succeeded Johnny Farrell as the club pro in 1972. Dave had been assistant pro at Baltusrol under Farrell both before and after World War II. And he had been the pro at three other clubs over a 20-year span, including Upper Montclair Country Club where he was stationed for 17 years.

In 1977 Carolan was succeeded as Baltusrol's pro by Bob Ross.

Bob Ross

An unusually well rounded and versatile golf professional — player, teacher, and statesman — Robert A. Ross, Baltusrol's pro at the time of this writing, has met the club's ever-expanding needs admirably.

A native of Vermont, Bob graduated from Admiral Billard Academy in New London, Connecticut, and Pasadena City College in California, and he attended PGA business schools and merchandise seminars. He was head golf professional at five other clubs before coming to Baltusrol in 1977, including the prestigious Philadelphia Cricket Club and Sawgrass Country Club in Ponte Vedra, Florida.

Bob Ross was part of the Professional Golf Association tour from 1959 to 1976, and he participated in three National PGA Championships and three U.S. Opens as well as winning both the Pennsylvania Open and Philadelphia PGA Championships in 1967. Ross was voted National Country Club Professional of 1980 by *Country Club Golfer Magazine* and the New Jersey Section of the PGA.

A former president of the Philadelphia Section of the PGA and a one-time member of the National PGA Rules Committee, Bob has also worked with many organizations concerned with the business of golf.

A remarkably attractive and personable gentleman, as well as a popular golf teacher, Bob has attracted able assistant professionals, and as a skilled merchandiser, he has run one of the largest and most beautiful pro shops in all of New Jersey.

Johnny Farrell

OTHER SPORTS AND ACTIVITIES

MEMBERS AND GUESTS

Other Sports and Activities

Although golf is the only sport currently played at Baltusrol, a number of other sports have appeared and exited over the years.

As early as 1901, the green committee was authorized to build two tennis courts. In 1907 additional (or new) tennis courts were planned for a newly acquired 25-acre tract. And after a sports committee had been reestablished in 1911 to handle tennis, squash, and trap shooting, the tennis courts were fixed up and were in heavy use by 1915.

The squash courts were located in the annex across from the front door of the club, which is now the ladies' locker room.

The tennis courts were flooded in the winter of 1915 and lighted at night for ice skating by the members.

Trap shooting had been instituted sometime before 1913, for it was in that year that $500 was spent for a new shooting platform and shelter and $350 for an additional trap. This sport probably was abandoned during the Depression, although it was considered again in 1935.

Fox hunts took off from the club grounds in the middle teens.

Bowling on the green was instituted in 1937 when the new green was completed near the eigh-teenth Upper golf tee. Night lights were installed, and an instructor engaged.

Trout fishing was enjoyed for some years by 20 members, more or less, who contributed about $10 to $15 per year to purchase rainbow trout and native trout to stock the pond, which is to the right of the thirteenth Upper fairway. The rules governing the sport were:

Fly fishing only.
Limit 6 fish per day.
Limit 12 fish per week.
No guests.
Gates in Turkey Lane (now Hillside Avenue) to be
    kept closed.
No fishing after dark.

There were frequent football outings to Princeton followed by dinner dances at the club during its middle and later years.

Lectures were held at the club, and dancing classes were conducted by the Arthur Murray Studio in the late 1930s.

And first-run movies were frequently shown on the lawn and the terrace during those years. They continued for many years.

In 1974 and 1975 paddle tennis was studied and voted against. The only nongolf sport that survived the years was bowling, which is conducted on public lanes outside the club.

The ultimate abandonment of all sports except for golf has made Baltusrol less attractive to families with children and no doubt has contributed to its difficulties in maintaining membership during its middle years. But it *is* a golf club and *solely* a golf club, and its steadfastness in maintaining that image has steadily enhanced its attractiveness to potential golfing members.

Watch Hunt — Thanksgiving 1913

1933 Ryder Cup Team

Members and Guests

Baltusrol's greatest story may lie in the singular accomplishments of its more distinguished members and guests over the years—the leaders in government, business, finance, the professions, and the arts who have found the club a congenial center for their social and recreational activities. The influence of these gifted people on national and international events and their overall contributions to the social order must be immense. They are the true movers and leaders in the broad socioeconomic community. And although snobbish overtones are frequently ascribed to clubs where such leaders congregate, the overtones are much less compelling than the basic accomplishments that brought them there—often from modest beginnings. For Baltusrol is not as much a citadel of wealth and power as it is a haven of accomplishment.

But once at Baltusrol, golf prevails—not tennis or squash or fishing or even skeet shooting or riding to the hounds, although all were once part of the club. Ultimately they gave way to golf—the all-pervasive interest that on a given Saturday morning teams up the old and the young, the executive and the novice—and on occasion, even the King and the commoner.

Johnny Farrell, Baltusrol's august pro, had met Edward, the Duke of Windsor, while on a Ryder Club team in England. When the Duke came to New York, he visited Johnny for playing lessons. He played only with Farrell or the club president, Bob Finney, on several occasions.

"The Duke had a good swing," Farrell said, "but his short game was poor because he was simply too tense." Nevertheless, at a White House dinner given in his honor by President Nixon, the Duke singled out Johnny, who was also present, as "the man who saved my golfing life."

During one of the Duke's last visits, I saw this pleasant old man with baggy knickers, an old sweater too long for him, and bags under his eyes, and I reflected on the Duke's exalted place in world affairs—beginning with his youthful hunting trips with Nicholas and Wilhelm early in the century, through World War I, up to his later withdrawal as head of the world's greatest empire.

I conveyed my excitement over this fleeting brush with greatness to my family that evening—including my son Jim.

Several weeks later, discussing English history, Jim's instructor at Seton Hall Prep reached the reign of Edward VIII. He paused to ask whether anyone knew the story of Edward.

Jim responded with a description of Edward's short reign, his marriage, and abdication.

"That's excellent, Jim," the instructor commented. "How did you come to know so much about Edward?"

"Oh," Jim replied, "he plays golf at Baltusrol."

The incredulous laugher that followed drowned any chance for anyone to explain that history is proximate, it is here and now—and can be found even at the local golf club.

Nixon himself became a member of Baltusrol when he moved from California to New York in 1965, but he resigned in August 1968 after the press attempted to make his membership a political issue. With a handicap of around 16, Nixon played pretty respectable golf—mostly with his law partners, John Alexander and Bob Guthrie.

Former President Gerald Ford played at the club while attending meetings at Beneficial Finance in Morristown.

Bill Eldridge, Baltusrol's long-time cocktail waiter, and Ernie Mudd, the maître d', spoke of the many celebrities they encountered during their long association with the club. Eldridge, a crack golfer and an accomplished pianist, was employed in 1946; and Mudd, a skilled archer, was employed in 1949. Mudd was a descendant of Dr. Samuel Alexander Mudd, the doctor who treated John Wilkes Booth after the assassination of President Lincoln and who was sentenced to prison for this humane act but was later pardoned, and who was honored 90 years later with a bronze memorial at Fort Jefferson commemorating his service during the yellow fever epidemic in 1867.

There were people from the entertainment world such as Douglas Fairbanks, Jr., Helen Hayes, Bing Crosby, Bob Hope, Gladys Swarthout, and Sylvia Sydney—her escort called her "Syd". There was Jimmy Dorsey, "a quiet person who loved to play golf here," and Sammy Kaye, "who drank four vodka and tonics on the terrace and then shot an 80."

There was Ferdie Grofé, composer of Grand Canyon Suite, who according to Mudd enjoyed sipping a cocktail in his wheelchair in the terrace room.

Noted athletes, of course showed up. Gene Tunney was a guest of Len Ulrope, who reportedly demonstrated his own pugilistic potential by bashing in the side of a locker with his bare fist. "Len, if I had a punch like that," Tunney is supposed to have said, "I'd never have left the ring."

The military was there too, including General Montgomery, commander of the Strategic Air Command in Korea, who was having lunch at the club when President Kennedy was assassinated.

There were many captains of industry and finance too who have not yet been mentioned in some other context. Heads of nationally known companies who come to mind are Carl Badenhausen, chief of Ballantine Brewery; Leslie Cassidy, chairman of Johns–Manville; Donald McGraw, chairman of McGraw-Hill; Orville Beal and Don MacNaughton chairmen of Prudential Insurance; William Bienecke of the Green Stamp Company; Lee Bickmore, chairman of Nabisco; William Black, managing partner of Peat, Marwick & Mitchell; and Walter Staub, managing partner of Coopers & Lybrand. The Staub family's club memberships spanned more than 50 years.

And there are Paul Stillman, chairman of both First National State Bank of New Jersey and Mutual Benefit Life Insurance Company; Robert Ferguson, also chairman of First National State Bank; and Robert Van Beuren, chairman of Midlantic National Bank. Also John Connor and Henry Gadsden, chairmen of Merck & Company. Connor was also chairman of Allied Corporation after serving for several years as Secretary of Commerce in President Lyndon Johnson's cabinet. Edward Hennesy, Connor's successor at Allied is also a club member. So is Paul Gorman, who was chairman of Western Electric, Penn Central, and International Paper. Paul had joined Penn Central to save it from disaster but quickly found it beyond redemption. He arranged with counsel one Sunday afternoon in 1970 to throw the company into bankruptcy, which current club president Paul Hanna found appalling because his bank had made sizable loans to Penn Central. The bank eventually recovered all.

Also David Foster, the chairman of Colgate, Palmolive, who helped host the Dinah Shore Ladies Golf Tournament on national TV.

At least four chairmen of Public Serivce Electric & Gas have been members: Donald Luce, Edward Snyder, Edward Eberle, and Robert Smith.

There is also Nestor MacDonald, head of Thomas & Betts, who in 1984 was still shooting his age (at 88)—and with a lame hip! Nestor was chairman of the Caddy Scholarship Foundation of the New Jersey State Golf Association during its greatest growth period, 1957–1967. Also, Vice Admiral Howard Orem and Rear Admiral Albert Mumma, who as chief of the Bureau of Ships was partly responsible for the SS *United States*—a luxury liner that was designed to be easily converted into a troop ship. Mumma was later president of Worthington Pump and, according to Chairman Walther Feldmann, was "the best engineer I ever met."

Wilson Newman was still another colorful fellow. A former chairman of Dun & Bradstreet, he was a member of the U.S. Price Control Board.

Then there was H. I. Romnes, chairman of AT&T, a modest and friendly man, and Jim Olson, currently vice chairman. Also Tom Bolger, chairman of Bell Atlantic, and Brad Wiley of John Wiley & Sons, publishers. Donald Regan, head of Merrill, Lynch and later Secretary of the Treasury, was also a member. So was John D. J. Moore, who left in 1969 to serve as U.S. ambassador to Ireland.

At the time of this writing, the board of governors included several top executives of well-known companies—for example: Paul Hanna, president of the club and chairman of Government Employees Financial Corporation; Mark Anton, president of Suburban Propane Gas; James Farley, president of Booz-Allen and Hamilton; Kenneth Nichols, CEO of Home Life Insurance; and Thomas Sayles, chairman of Summit Bankcorp.

Also there is Joe Partenheimer, retired head of Atlas Supply Company, a joint venture of five Standard Oil Companies. Joe, who could never say no to an appeal for help, was, at 88, still working with blacks and whites in Newark and trusted implicitly by both. As a member of the board of governors, he initiated the *Baltusrol News* in 1961 and was its first editor. An active golfer, Joe was named Baltusrol Sportsman of the Year in 1983.

Harry Taylor, still an active golfer at age 90 in 1985, was captain of the East Orange High School football team early in the century and graduated from Colgate in the teens before entering his father's real estate business in 1920. Harry initiated several large shopping centers and other large developments. He ultimately became the dean of state realtors as well as chairman of Baltusrol's real estate committee during his 1956–1960 term on the board of governors.

And Fred Herrigel, Jr., long-time head of the prominent Millburn law firm bearing his name, was also still an active golfer in 1985 at age 90.

There are, of course, other distinguished members whose accomplishments no doubt have equaled those mentioned. But the names cited convey something of the flavor of the membership.

One final distinction: Fred Rummel, who joined the club in 1939, was brought up in Pottsville, Pennsylvania, with John O'Hara. Indeed, John's father was the Rummels' family doctor. And Fred was a character in O'Hara's first and probably best-known book *Appointment in Samara*. It was a thinly disguised story of actual people and events in Pottsville, which O'Hara renamed "Gibbsville." Fred, who can be identified as the young representative of City Bank in the area, was the rejected suitor of Caroline Walker, who ultimately married the story's tragic principal character, Julian English.

Fortunately, Mae Rummel did not reject this courtly man. Mae was women's golf chairperson at Baltusrol in 1945 and 1946.

INDECISION AND HEADACHES

The Baltusrol board has had many governors over the years who were steeped in finance, engineering and construction, and general management knowledge, all of which enabled it to make major decisions quickly. But the board seemed to have great trouble resolving the *minor* issues. Here are some examples.

Golf Carts

In January 1956 the green committee opposed the acquisition of electric carts for use by members, and the board of governors formally rejected them in April 1957. Thereupon the following occurred:

October 29, 1957	—Board approved purchase of one golf cart for green personnel to determine whether carts damage the golf course.
October 18, 1958	—Golf carts under study.
January 27, 1959	—"Few would want golf carts under the health restrictions suggested."
April 26, 1960	—"No carts."
June 26, 1962	—Cart issue to be reconsidered.
July 24, 1962	—Bought 6 carts. Heavy requirements for their use—doctor's certificate, etc.
November 27, 1962	—Women want to be permitted to use golf carts.
April 23, 1963	—Golf carts restrictions eased a little.
May 26, 1963	—Voted no golf carts for guests or women.
July 28, 1964	—Member may take two carts for guests.
December 22, 1964	—Estimated net income from golf carts for year $3,910—authorized leasing of 12 carts for use on both courses.
January 25, 1966	—Member's wife giving him a golf cart for Christmas—can he use it? No!
March 26, 1968	—Golf carts can be used by members over age 70 on written request. Others on first-come, first-served basis.

Thereafter, all restrictions on the use of carts were removed and many carts were leased and later bought.

Shorts

In August 1950 Secretary Finney of the board of governors reminded the members that shorts were not to be worn by men or women, that women could not wear slacks, and that men must wear shirts. Thereafter, the following board actions occurred:

May 25, 1954	—No shorts permitted for men or women.
May 24, 1960	—No shorts to be worn in 1961 USGA Women's Open to be held at the club—club rule.
February 28, 1961	—Shorts authorized for 1961 USGA Women's Open only.
March 28, 1961	—Members vote 229–175 in favor of wearing shorts.
May 25, 1961	—Resolution to continue ban on wearing shorts tabled.
July 23, 1968	—Governors vote against shorts, 12 to 4.
July 22, 1969	—Women: no shorts, slacks, or pants suits. Men: no shorts.
February 4, 1970	—Women may wear slacks from November 1 to April 1.
October 27, 1970	—Women may wear pants suits inside the club.
April 12, 1971	—Slacks for women after April 1 suggested. No action taken.
September 12, 1971	—Infestation of mosquitoes. Women may wear slacks.
October 22, 1972	—No shorts for men.

What might have been

Headaches

During the almost 90 years of the club, the board of governors has had to deal with many nettlesome problems and headaches. Here are a few:

Poor condition of fairways	Intermittently—usually in August
Slow play	Constantly—1895 to date
Too many association or "trade" tournaments	Intermittently—1895 to date
Abuse of guest privileges	Intermittently—1895 to date
Tipping—restaurant employees	Intermittently—1895 to date
Members buying golf balls from "gypsies" intead of pro shop	Early
Playing cards and dice for drinks	Around 1936
Prohibition	1920 to 1933
Women golfers said to be taking over	Intermittently—1895 to date
Women golfers bringing dogs into locker room	Around 1937
Female entertainers banned from clubhouse—policy waived for Stewart Baker dinner in 1938	Around 1932–1938
Poor township maintenance of Baltusrol Way	Pre-1925
Indecision concerning where club is located—Short Hills or Springfield	1895 to 1935
Caddies sitting around front door	1901
Caddies sitting on putting green fence	Intermittently
Caddie strike	1941
Disappearing bookkeeper—$2,297 missing	1923
Vandalism by kids in the area	Constantly
Raucus roadhouse opposite Shunpike entrance	1925
Peeping Tom	October 26, 1900
Greenkeeper's farmhouse office entered twice despite big mongrel dog	January 24, 1960
Caddie runs cart into tree	1963
Chandelier falls in East Lounge	1963
Houseman gets hernia lifting beer barrel	1963
Two diamond rings stolen from member—former houseman caught but had "lost" the rings	1963
Leak in terrace floor	1956–1985
Need for a club history	1956–1985

APPENDIX

Presidents

Arthur D. Weeks	1895–1898	Monroe J. Rathbone	1955–1958
Louis P. Bayard	1898–1919	John C. Smaltz	1958–1961
Robert S. Sinclair	1919–1926	Walther H. Feldmann	1961–1964
William G. McKnight	1926–1929	William M. Walther	1964–1967
J. Stewart Baker	1929–1938	Robert Finney	1967–1970
Caxton Brown	1938–1943	Matthew J. Glennon	1970–1973
Walter R. Hine	1943–1945	John S. Roberts	1973–1976
Maurice N. Trainer	1945–1947	B. P. Russell	1976–1979
Stoddard M. Stevens	1947–1952	Robert J. Boutillier	1979–1982
Hobart C. Ramsey	1952–1955	Paul J. Hanna	1982–

Club Champions

1899–1900	James A. Tyng	1947	W. G. Johnston
1901–1902	R. E. DeRaismes	1948–1949	L. E. Carpenter, Jr.
1903–	F. C. Reinhart	1950	H. O. Lichtenwalter, Jr.
1904	L. P. Bayard, Jr.	1951	George H. Frey
1905	A. S. Morrow	1952	W. G. Johnston
1906	Walter Kobbe	1953	Stephen G. Lee
1907–1908	C. J. Sullivan	1954	L. E. Carpenter, Jr.
1909	Max H. Behr	1955	Richard A. Henry
1910	Oscar Woodward	1956	L. E. Carpenter, Jr.
1911	C. J. Sullivan	1957	H. O. Lichtenwalter, Jr.
1912	C. E. Vanvleck, Jr.	1958–1960	L. E. Carpenter, Jr.
1913	W. F. Morgan, Jr.	1961–1962	H. O. Lichtenwalter, Jr.
1914–1916	Max Marston	1963	David Baldwin
1917	Stephen P. Nash	1964	John Farrell, Jr.
1918	R. H. Gwaltney	1965	John Poinier
1919	A. F. Kammer	1966	Oliver H. Havens
1920	S. Van Vechten	1967–1968	Edward B. Crosland, Jr.
1921	A. F. Kammer	1969	John Farrell, Jr.
1922–1929	E. M. Wild	1970–1971	John R. Schermerhorn, Jr.
1930	A. F. Kammer, Jr.	1972	John Farrell, Jr.
1931	E. M. Wild	1973	L. E. Carpenter, Jr.
1932	F. Jefferson, Jr.	1974	R. T. Gaertner
1933	Robert Finney	1975	Jerry Swon
1934–1935	M. S. Lindgrove	1976–1977	R. T. Gaertner
1936–1938	A. F. Kammer, Jr.	1978	Christopher B. Scott
1939–1940	M. S. Lindgrove	1979	Eric J. Gleacher
1941	H. S. Hall	1980	L. Randy Riley
1942	Rollin P. Taylor	1981	John Farrell, Jr.
1943	P. H. Hartung	1982	Gilbert A. Zimmerman, Jr.
1944	Richard A. Henry	1983	Gilbert A. Zimmerman, Jr.
1945	Eugene F. Krautter	1984	Bryan Burke
1946	Richard A. Henry		

Seniors Association Champions

1922	Robert Sinclair	1958	George A. Murray
1923–1932	R. C. James	1959	George H. Frey
1933	C. N. Fowler	1960	Robert Finney
1934	J. R. Monroe	1961	Robert Finney
1935	A. F. Kammer	1962	R. A. Henry
1936	H. N. Balch	1963	A. Konkle
1937	John F. Duffey	1964	Robert Finney
1938	Caxton Brown	1965	Robert Finney
1939	H. N. Balch	1966	Thorwald Larson
1940	M. S. Lindgrove	1967	R. A. Henry
1941	M. S. Lindgrove	1968	Robert Finney
1942	C. E. Dwyer	1969	William Y. Dear, Jr.
1943	W. P. Conway	1970	William Y. Dear, Jr.
1944	Arthur E. Jones	1971	John Roberts
1945	E. J. Ogden	1972	John Poinier
1946	M. S. Lindgrove	1973	John Poinier
1947	W. P. Conway	1974	Thorwald Larson
1948	W. V. Cadmus	1975	John A. Deitrich
1949	E. F. Krautter	1976	W. W. Reinhard
1950	S. R. Williams	1977	John A. Deitrich
1951	M. S. Lindgrove	1978	Thorwald Larson
1952	M. S. Lindgrove	1979	John A. Deitrich
1953	M. S. Lindgrove	1980	John A. Deitrich
1954	S. R. Williams	1981	Arthur R. Paterson
1955	W. W. Krautter	1982	John A. Deitrich
1956	R. A. Henry	1983	John A. Deitrich
1957	George H. Frey	1984	L. E. Carpenter, Jr.

Outside Championships

New Jersey Senior Championship

1928—R. C. James
1940–1941, 1944–1945—A. F. Kammer
1947–1950, 1952–1954, 1957—M. S. Lindgrove
1963—R. A. Henry
1967—Thorwald Larson
1970, 1971—John Poinier
1976—William Y. Dear, Jr.

U.S. Senior Championship

1910—Frank A. Wright
1912–1915—James A. Tyng
1914—Frank A. Wright
1947—M. S. Lindgrove

International Senior Championship

1951—M. S. Lindgrove

New Jersey Amateur Championship

1924–1926—A. F. Kammer

New Jersey Open Championship

1954—David Baldwin

Women Champions

Year	Champion	Year	Champion
1916	Mrs. S. P. Nash	1954	C. Helen Hockenjos
1919	Elizabeth A. Hardin	1955	Charlotte DeCozen
1920–1921	Mrs. S. F. Dubois	1956	C. Helen Hockenjos
1922	Elizabeth A. Hardin	1957	Charlotte DeCozen
1923	Elizabeth A. Hardin	1958	C. Helen Hockenjos
1929	Mrs. R. F. Decker	1959	C. Helen Hockenjos
1930	Mrs. C. L. Voorhees	1960	Helen R. Hawes
1931	Charlotte E. Clutting	1961	Essene D. Baldwin
1932	Mrs. C. L. Voorhees	1962	Lucille Hickman
1933	Mrs. C. L. Voorhees	1963–1968	C. Helen Hockenjos
1934	Mrs. C. H. Donner	1969	Lucille Hickman
1935	Alice Rutherfurd	1970	Isabel Mercer
1936	Mrs. W. D. Goss, Jr.	1971	Isabel Mercer
1937	Alice Rutherfurd	1972	Brita Blauvelt
1938	Alice Rutherfurd	1973	Harriet Hart
1939	Gail Wild	1974	Ruth Swart
1940	Gail Wild	1975	Isabel Mercer
1941	Mrs. Bruce Ryan	1976	Nancy Rogers
1942	Gail Wild	1977	Betty Lou McCabe
1943	Mrs. H. C. Ramsey	1978	Dorothy Paluck
1944–1946	Mrs. H. C. Ramsey	1979	Ruth Swart
1947	Helen R. Hawes	1980	Janet Wilcox
1948	C. Helen Hockenjos	1981	Dorothy Paluck
1949–1950	Charlotte DeCozen	1982–1983	Robin Jervey
1951–1952	C. Helen Hockenjos	1984	Lee Steffens
1953	Charlotte DeCozen		

Women's Golf Chairpersons

Name	Years	Name	Years
Mae Rummel	1945–1946	Mary Lou Hodge	1967–1968
Ruth Evers	1947–1948	Marie Tansey	1969–1970
Helen Hawes	1949–1950	Ruth Swart	1971–1972
Peg Pelletier	1951–1952	Bette Gaertner	1973–1974
Claire Beckett	1953–1954	Kay Ellison	1975–1976
Jessie Hassell	1955–1956	Nancy Rogers	1977–1978
Janice Vilett	1957–1958	Carmen Mumma	1979–1980
Ernestine Murray	1959–1960	Isabel Mercer	1981–1982
Ruth Lichtenwalter	1961–1962	Janet Talbot	1983–1984
Elizabeth Johnston	1963–1964	Marjorie Reinhard	1985–1986
Helen Kalchthaler	1965–1966		

USGA Championships Held at Baltusrol

Open Championships

Year	Entrants	Winning Score	
1903	89	307	Tie—Playoff won by Willie Anderson 82; second David Brown 84.
1915	141	297	Won by Jerome D. Travers (Amateur); second Tom McNamara.
1936	1,277	282	Won by Tony Manero; second Harry E. Cooper.
1954	1,928	284	Won by Ed Furgol; second Gene Littler.
1967	2,651	275*	Won by Jack Nicklaus; second Arnold Palmer.
1980	4,812	272*	Won by Jack Nicklaus; second Isao Aoki.

*New record

Amateur Championships

Year	Entrants	
1904	142	H. Chandler Egan defeated Fred Herreshoff 8 and 6. Medal won by H. Chandler Egan 242 (54 holes).
1926	157	George Von Elm defeated Robert T. Jones, Jr., 2 and 1. Medal won by Robert T. Jones, Jr., 143 (36 holes).
1946	899	Stanley E. Bishop defeated "Smiley" L. Quick 1 up (37 holes). Medal won by "Skee" Riegel 136 (36 holes—a new record).

Women's Open Championship

Year	Entrants	Winning Score	
1961	85	293	Won by Miss Mickey Wright, second Miss Betsy Rawls.

Women's Amateur Championships

Year	Entrants	
1901	84	Miss Genevieve Hecker defeated Miss Lucy Herron 5 and 3. Four players were tied for the medal with scores of 97.
1911	67	Miss Margaret Curtis defeated Miss Lillian B. Hyde 5 and 3. Medal won by Mrs. Ronald H. Barlow with a score of 87.

INDEX

Alexander, John, 144
Amory, Cleveland, 15
Anderson, Norb, 113
Anderson, Willie, **18**, 19, 22, 41,
    44, 98, 100, **100**, 114, 137, 154
Anton, Mark, 145
Aoki, Isao, 40, 114, 116, 154
Armitage, Mrs. Clifford, 130

Badenhausen, Carl, 145
Baker, George, 42, **42**
Baker, J. Stewart, 27, **28**, 29, 31,
    149, 152
Baker, Marianne, 31
Balch, H. N., 152
Baldwin, David M., 126, **127**, 152,
    153
Baldwin, Mrs. Donald M., 130
Baldwin, Essene D., 153
Baldwin, Lycidias, 11, 12
Baltusrol Golf Club
    beginning of, 14
    conception, 14
    founders, 5, 14-18
    location, 14, 61
    site, historical, 9-10
Baltusrol Way, 25, **25**, 44, 57, 149
Barlow, Mrs. R. H., 102, 155
Barnes, Long Jim, 103
Barnes, Shepard, 34
Bassett, Carroll P., 17
Bassett, Charles, 17
Bayard, L. P., 17, 19, 26, 29, 123,
    152
Bayard, L. P., Jr., 152
Beal, Orville, 145
Beckett, Claire, 154
Beekman, William B., 17
Beekman, Mrs. William B., 20
Behr, Max, 118, 152
Bell, Charles L., 21
Beman, Deane, 113
Berg, Patty, 133
Bickmore, Lee, 145
Bienecke, William, 145
Bigelow, Miss, 17
Bishop, Georgianne, 19
Bishop, Stanley E. ("Ted"), 34, 106,
    106, 154
Black, Pearl, 130
Black, William, 145
Blaubelt, Mrs. Reginald T., Jr.
    (Brita), 130, 154
Board of governors, 18, 26, 27, 36,
    39, 42, 126, 128, 145, 148-149
    first, 14
Boissevain, Daniel C., 17
Bolger, Tom, 145
Booth, John Wilkes, 144
Boros, Julius, 107
Boutillier, Robert J., 40, 41, **69**, 152

Brown, Caxton, 27, **28**, 31-32, **136**,
    152
Brown, David, 19, 100, 154
Brown, Elaine, *see* Mole, Mrs.
    Harvey E.
Brush, John, 20
Burgess, C. P., 35, 60, 124
Burgess, Newton A., 124
Burke, Bryan, 152
Burke, Jackie, 107
Burns, Tom, 33, 36

Caddies, 25, 149
Cadmus, W. V., 153
Cahill, Bernard, 20
Cain, Joe, 11
Calhoun, Jesse, 11
Campbell, Bill, 109
Campbell, Willie, 19
Carolan, David P., 139
Carpenter, Lawrence E., Jr., 122,
    126, **127**, 152, 153
Casey, Ed, 33, 37, 42, 62, **63**
Casper, Billy, 98, 112
Cassidy, Leslie, 145
Chandler, Arthur B., 22
Chapman, Dick, 110
Chazournes, Heloise de, 15
Cleveland, Grover, 14
Clubhouse
    fire (March 27, 1909), 20-21
    new, 21, 33
    old, **15**, 16, **17**, 20-21
    renovations, 41
Clutting, Charlotte E., 153
Cocktail Lounge, **73**
Collins, James, 5
Connell, Fred, 123
Connor, John, 145
Conway, William Palen, 30, 34, 152,
    153
Cooper, "Lighthorse" Harry, 105,
    154
Crosby, Bing, 144
Crosland, Edward B., Jr., 152
Cudone, Caroline, 133
Curtis, Harriot, 102
Curtis, Margaret, 19, 24, 45, 98, 99,
    102, **102**, 155
Cushing, Charles Phelps, 16

Davis, Peter B., 11-12
Day, Joseph P., 26, 27
Dear, William Y., Jr., 126, **127**, 128,
    153
DeBary, Miss L., 129
DeBois, James, 5
Decker, Mrs. B. F., 153
DeCozen, Charlotte, 130, 153
Demaret, Jimmy, 107
De Noble, Mark, 5, 41, 42, **42**
DeRaismes, R. E., 152

Dietrich, John A., 122, 128, **128**,
    153
Dining Room, **74**
Donner, Mrs. C. H., 153
Dorfman, Gary, 138
Dorsey, Jimmy, 144
Dubois, Mrs. S. F., 153
Dudley, Ed, 30
DuFais, John, 14-15, 17
Duffey, John F., 152
Dunn, Willie, 19
Durnien, Frank, 42, **42**
Dwyer, C. E., 152

East Lounge, **75**
Eaton, Zell, 110
Eberle, Edward, 145
Edison, Thomas, 20
Edward, Duke of Windsor, 10, 139,
    144
Edward VII, King (England), 10
Egan, H. Chandler, 19, 44, 98, 101,
    101, 154
Eisenhower, Dwight D., 35
Eldridge, William, 5, 42, **42**, 144
Elliott, Len, 5
Ellison, Kay, 154
Evans, Charles ("Chick"), Jr., 45,
    98, 103, 104, 105, 106, 123
Evers, Ruth, 154

Fairbanks, Douglas, Jr., 144
Farley, James, 145
Farrell, Billy, 139
Farrell, Cathy, *see* Rock, Mrs. Cathy
Farrell, Jimmy, 138
Farrell, Johnny, 5, 30, **30**, 31, **35**,
    42, 60, 98, 100, 103, 106, 136,
    137, 138-139, **139**, 144
Farrell, Johnny, Jr., 5, 40, 122, 128,
    128, 138, 152
Farrell, Kay, **35**, 138
Farrell, Peggy, *see* McGuire, Mrs.
    Peggy
Feldman, Walther H., 39, 145, 152
Fergus, Keith, 114
Ferguson, Robert, 145
Finney, Robert, **38**, 39, 44, 57, 58,
    62, 122, 124-125, 144, 148, 152,
    153
Fire, clubhouse (March 27, 1909),
    20-21
Fisher, John W., 106
Flaherty, Joseph, 42, 62, **62**
Fleckman, Marty, 112
Fleming, Thomas, 5, 10
Ford, Doug, 107
Ford, Gerald, 144
Forrester, Jack, 137
Foster, David, 145
Fowler, C. N., 152
Franklin, William, 9

Frey, George H., 152, 153
Furgol, Ed, 36, **107**, 107-109, 113, 154

Gadsden, Henry, 145
Gaertner, Bette, 130, 154
Gaertner, R. T., 152
Garage, 21, 33
Garden State Women's Golf Association, 130
Gershwin, George, 31
Ghezzi, Vic, 105
Gilbert, Cass, 14
Gleacher, Eric J., 152
Glennon, Matthew J., 39, 152
Goalby, Bob, 112
Goodman, Johnny, 98, 106
Gorman, Paul, 145
Goss, Mrs. W. D., Jr., 129, 130, 153
Gourley, Tom, 136, 137
Graham, L. E., 19
Grainger, Ike, 110
Great Depression, 27, 29
Green, Tom, 123
Griscom, Frances C., 19, 98, 99
Grofé, Ferdie, 144
Grogan, Allen, 61-62
Guthrie, Bob, 144

Hagen, Walter, 98
Hall, H. S., 152
Hanna, Paul J., 5, 41, **41**, 145, 152
Hannigan, Frank, 58, 98
Hardin, Elizabeth A., 130, 153
Harding, Warren G., 137
Harmon, Claude, 112
Hart, Mrs. A. Sherburne (Harriet), 130, **132**, 133-134, 154
Hartshorn, Stewart, Jr., **26**
Hartung, Philip, 34, 124, 152
Hassell, Jessie, 154
Hauser, Johnny, 31
Havemeyer, Fred, 123
Havemeyer, Harry, 123
Havemeyer, Theodore, 123
Havens, Oliver H., 152
Hawes, Mrs. Theodore W. (Helen R.), 129, 130-131, **131**, 153, 154
Hayes, Helen, 144
Hayes, Mark, 114
Headquarters, 26
Hecker, Genevieve, 19, 44, **99**, 129, 155
Hennesy, Edward, 145
Henry, Richard A., 122, **125**, 126, 152, 153
Herd, Fred, 22
Herreshoff, Fred, 101, 154
Herrigel, Fred, Jr., 145
Herron, Lucy, 99, 155
Hickman, Lucille, 154

"Hill Property Development Project," 40
Hilton, Harold, 22, **23**, 98
Hine, Walter R., 32, 33, 34, 152
Hinkle, Lon, 114
Hoag, Bob, 112
Hockenjos, C. Helen, 130, 131, **132**, 133, 153, 154
Hodge, Mary Lou, 154
Hoffman, Red, 58, 98
Hogan, Ben, 19, 31, 39, 41, 98, 100, 104, 107, 109, 110, 112, 113, 114
Hope, Bob, 144
Hoyt, Beatrix, 15, 19, 98
Hoyt, Henry, 31
Hunter, David, 137
Hunter, George, 16
Hyde, Lillian, 102, 155
Irwin, Hale, 98, 114
Isenberg, David, 138

Jackson, Andrew, 11
Jacobus, George, 119
James, R. C., 122, 124, 152, 153
Jameston, Betty, 133
Jefferson, F., Jr., 152
Jehlen, Carl J., 5, 27, 31, **36**, 36-37, 39, 41, 42
Jehlen, Marie, 41
Jepson, Helen, 31
Jervey, Robin, 131, **131**, 154
Jessen, Ruth, 111
Johnston, Elizabeth, 154
Johnston, W. G., 152
Jones, Arthur E., 153
Jones, Bobby, 19, 27, 30, **30**, 41, 61, 98, 100, 103, **104**, 123, 138, 154
Jones, R. Avery, 27, 29, 32, 33, 42, 62, 105, 137
Jones, Robert Trent, 34, 35, 59-60, 114

Kalchthaler, Helen, 154
Kammer, A. F., 60, 118, 122, 123-124, 152, 153
Kammer, A. F., Jr., 106, 123, 152
Kaye, Sammy, 144
Kean, John, 22
Kean, Thomas, **41**
Keller, Charles M., 15
Keller, Louis, 5, 14-18, 19, 20, 24-25, 26, 27, 31, 42, 44, 58, 100, 124
Kessler, Kaye, 98
Kirk, Chester H., 21
Kirkby, Oswald, 118
Knowles, Ellis, 124
Kobbe, George C., 17
Kobbe, Walter, 152
Konkle, A., 153

Kosky, Pete, 137
Krautter, Eugene F., 152, 153
Krautter, W. W., 153

Larned, William A., 17
Larson, Thorwald ("Tom"), 122, 128, 153
Lauder, Harry, **137**
Lee, Stephen G., 152
Leuders, Frederick J., 40, 44
Leuders, Mrs. Frederick J., 40
Lewis, Mrs. Albert Nelson, 16
Lichtenwalter, Mrs. Homer, 130
Lichtenwalter, Homer O., Jr., 122, **125**, 125-126, 152
Lichtenwalter, Ruth, 154
Lindgrove, Martin S., 122, 124, **125**, 152, 153
Littler, Gene, 98, **108**, 109, 154
Locke, Bobby, 107
Low, George, 22, **23**, 45, 136, **136**, 137, **137**
Lower course, 25, 58, 59-61, **68-71**, **76, 78-79, 82, 84, 88-91, 95-96**
Luce, Donald, 145

MacDonald, Bob, 103
Macdonald, Charles B., 123
MacDonald, Nestor, 145
Macfarlane, Willie, 138
MacNaughton, Don, 145
Mahon, James, III, 144
Maintenance, 61-62
Manero, Tony, 31, 105, **105**, 154
Mangrum, Lloyd, 107
Mangrum, Ray, 105
Manice, Mrs. Edward A., 129
Marston, Max, 98, 106, 122, 123, 152
Martin, H. B., 18, 29, 101
May, John, 20
Mayor, Dick, 106, **108**, 109
McAllister, Louise, 16
McAllister, Ward, 16
McBee, Rives, 112
McCabe, Betty Lou, 154
McCawley, W. W., 18
McGowan, Pat, 114
McGraw, Donald, 145
McGraw, John, 14
McGuire, Mrs. Peggy, 138-139
McKnight, William G., 27, 152
McNamara, Tom, 103, 154
Men's grill, 21, 27
Men's locker room, **75**
Members, honorary, 31, 34, 37, 123, 124, 137, 139
Membership, 16-17, 18, 25, 26, 27, 29-30, 32, 34, 37, 40, 41, 142
original, 16
Mercer, Isabel, 130, 154

Metropolitan Golf Association, 15, 34, 59
Middlecoff, Cary, 98, 106, 107
Miegoc, Joe, 98
Miller, Johnny, 98
Miller, Violet, 17
Molé, Harvey E., 32
Molé, Mrs. Harvey E., 32
Monroe, J. R., 31, 152
Moore, Mrs. Clement C., 17
Moore, John D., Jr., 145
Morgan, William Fellowes, 17
Morgan, Mrs. William Fellowes, 129
Morgan, William Fellowes, Jr., 152
Moriarty, Mary, 20
Mortgage, burning of the, 31, **32**
Morrow, A. S., 152
Mudd, Ernest, 5, 36, 42, **42**, 124, 144
Mudd, Rose, 124
Mudd, Dr. Samuel Alexander, 144
Mumma, Albert, 145
Mumma, Carmen, 154
Murray, Ernestine, 154
Murray, George A., 153
Murray, Mrs. George A., 130

Nash, S. P., 152
Nash, Mrs. S. P., 153
Nelson, Byron, 98, 119, 138
New Jersey Golf Association, 19
Newman, Wilson, 145
Nichols, Kenneth, 145
Nicklaus, Jack, 19, 31, 38, 40, 98, 100, 104, 112-113, **113**, 114, **115**, 116, 154
Nixon, Richard M., 139, 144
Noer, O. J., 61
Northbourne, Lord, 101
Norton, Willie, 19

O'Connor, Maurrie, 119
Ogden, Miss E., 17
Ogden, E. J., 153
O'Hara, John, 146
Old Course, 18, 24, 44, **44-57**, 58
Oliver, Miss M., 129
Olson, Jim, 145
Orcutt, Maureen, 131, 133
Orem, Howard, 145
Original golf course, 44
Ouimet, Francis, 24, 34, 98, 103, 104, 123

Palmer, Arnold, 38, 60, 98, 104, **112**, 112-113, **116**, 154
Paluck, Dorothy, 154
Park, Willie, Jr., 19
Parking lot, **22**, 34, 35, 39
Parks, Sam, Jr., 105
Partenheimer, Joseph, 5, 145
Paterson, Arthur R., 153

Patterson, Mrs. Myra D., 129
Patton, Billy Joe, 110
Pelletier, Peg, 154
Player, Gary, 98, 138
Poinier, John, 122, 128, 152, 153
Potter, Robert, 5, 62, **62**
Prentice, Jo-Anne, 111
Price, Charles, 98
Pro shop, 40
Professional Golfers Association of America, 58

Quick, Smiley L., 34, 106, 154

Rahway Valley Railroad, **19**, 20, **20**, **21**
Ramona, 31
Ramsey, Hobart C., 34, 35, 35-36, 37, 133, 152
Ramsey, Mrs. Hobart C. (Collette), 35, **35**, 130, **132**, 133, 153
Rathbone, Monroe J., 37, 152
Rawls, Betsy, 98, 111, 154
Ray Edward ("Ted"), 22, **23**, 24, 98, 103, 138
Reekie, Bill, 124
Regan, Donald, 145
Reid, Mike, 114
Reinhard, Marjorie, 154
Reinhart, F. C., 152
Renwick, W. W., 26
Riegel, Robert H. ("Skee"), 106, 154
Riley, L. Randy, 152
Roberts, John, 39, 40, 152, 153
Rock, Mrs. Cathy, 139
Rogers, Nancy, 5, 130, 133, 154
Roll, Baltus, murder of, 11-12
Roll, Mrs. Baltus, 11
Romnes, H. L., 145
Ross, Bob, 40, **69**, **74**, 139
Rudolph, Mason, 112
Rummel, Fred, 146
Rummel, Mae, 146, 154
Runyan, Paul, 105
Russell, B. P., 40, **41**, 152
Rutherford, Alice, 130, 153
Ryan, Mrs. Bruce, 153
Ryan, Mrs. Frederick B., 129, 130

Sarazen, Gene, 98, 105
Sayles, Thomas, 145
Sayre, Brook, 11
Schermerhorn, John R., Jr., 152
Scott, Christopher B., 152
Seely, Charles H., 118
Sharkey, Jack, 124
Sigel, Helen, 133
Simpson, Charles, 40, **69**
Sinclair, Robert S., 26, **26**, 27, 152
Sinclair, Wallace, 118
Smaltz, John C., 15, 39, 152

Smith, Alex, 22, **23**, 98, **100**, 137
Smith, Robert, 145
Smith, Willie, 19, 22, 137
Snead, Sam, 98, 107, 109-110, **110**
Snyder, Arthur, 40
Snyder, Edward, 145
Social Register, 15, 26
Springfield, Battle of, 9-10
Stackhouse, Mrs. Frank (Mary), 130
Starett, Paul, 22
Staub, Walter, 145
Steffens, Miss Lee, 130, 131, **131**, 154
Stevens, Stoddard M., 34-35, 152
Stillman, Paul, 145
Storey, Jean, 42
Stout, Mrs. Charles, *see* Hecker, Genevieve
Stranahan, Frank, 106
Sullivan, C. J., 118, 152
Sullivan, Des, 98
Sullivan, Ed, **35**
Swart, John, 5
Swart, Mrs, John (Ruth), 5, 130, 154
Swarthout, Gladys, 144
Sweetser, Jess W., 123
Swon, Jerry, 152
Sydney, Sylvia, 144

Taft, William Howard, 22, 137
Talbot, Janet, 154
Tansey, Marie, 154
Taylor, Harry, 145
Taylor, Rollin P., 152
Tellier, Louis, 103
Tennanbaum, Jack, 20
Terrace Lounge, **73**
Tillinghast, Albert W., 5, 25, **58**, 58-61, 101
Toler, Henry Pennington, 17, 20, 21, 123
Topping, Henry, 60
Trainer, Maurice N., 34, 152
Trainer, Mrs. Maurice N., 31
Travers, Jerome D. ("Jerry"), 24, 45, 98, 101, **101**, **103**, 103, 118, 154
Travis, Walter J., 98, 101, **101**, 118
Trees, 62
Trevino, Lee, 39, 98, 112, 113
Tucker, Willie, 19
Tunney, Gene, 144
Turf maintenance, 61-62
Turnesa, Willie, 106, 133
Turnure, Arthur B., 14, 15, 16, 17
Turnure, Mrs. Arthur B., 15
Tyng, James A., 122, 123, 152, 153

Ulrope, Len, 144
Underhill, Ruth, 98
Upper course, 25, 58, 59, 62, **77**, **80-81**, **85-87**, **92**, 133

Van Beuren, Robert, 145
Van Vechten, S., 152
Vanvleck, C. E., Jr., 152
Vardon, Harry, 22, **23**, 24, 98, 103, 138
Vare, Glenna Collett, 131, 133
Verdi, Jean, 36, 42, **74**
Vilett, Janice, 154
Von Elm, George, 27, 104, **104**, 123, 154
Voorhees, Mrs. C. L., 153
Walther, William M., 38, **38**, 39, 152
Ward, Bud, 106
Washington, George, 9, 10
Watson, Tom, 22, 98, 114, **116**
Weekes, Alice, 14

Weeks, Arthur D., 14, 17, 19, 152
Weiskopf, Tom, 114
Weitzel, Johnny, 110
Whiteman, Paul, 31
Wilcox, Janet, 154
Wild, E. M., 104, 118, **122**, 123, 124, 130, 152
Wild, Gail, 123, 130, 153
Wiley, Brad, 145
William, S. R., 153
Wilmerding, Mrs. John C., 17
Windler, Herbert, 100
Windsor, Duke of, *see* Edward, Duke of Windsor
Wingham, H. G., 123
Women's Golf Committee, 134
Women's Metropolitan Golf Association, 17, 129, 130
Women's New Jersey Golf Association, 130
Wood, Craig, 98
Wood, J. Walter, 17
Wood, Mrs. Kenneth, 130
Woodward, Oscar, 152
Wright, Frank A., 153
Wright, Miss Mickey, 18, 38, 98, 111, **111**, 154

Zaharias, Babe Didrikson, 18, 111, 133
Zimmerman, Gilbert A., Jr., 152
Zizi, Crossmier, 20

The Upper Course

Hole	Par	Yards	Hole	Par	Yards
1	5	463	10	3	151
2	4	422	11	5	551
3	3	191	12	4	333
4	4	384	13	4	363
5	4	390	14	4	380
6	4	428	15	3	139
7	3	214	16	4	435
8	5	533	17	5	557
9	4	321	18	4	459
Out	36	3346	In	36	3368
			Total	72	6714